Practical Tasks and Revision Exercises

Electrical Installation Series – Intermediate Course

E.G. Stocks, J.N.Hooper, M.Doughton and C.Duncan

2 Edited by Chris Cox

2

Australia · Canada · Mexico · Singapore · Spain · United Kingdom · United States

Practical Tasks and Revision Exercises

For more information, contact Thomson Learning, Berkshire House, 168–173 High Holborn, London, WC1V 7AA or visit us on the World Wide Web at: http://www.thomsonlearning.co.uk

---

British Library Cataloguing-in-Publication Data
A catalogue record for this book is available from the British Library

---

ISBN 1-86152-716-0

First published 2001 by Thomson Learning

Printed in Coatia by Zrinski d.d.

About this book

"Practical Tasks and Revision Exercises" is one of a series of books published by Thomson Learning related to Electrical Installation Work. The series may be used to form part of a recognised course or individual books can be used to update knowledge within particular subject areas. A complete list of titles in the series is given below.

"Practical Tasks and Revision Exercises" summarises the underpinning background knowledge required for the City and Guilds Course 2360 Part 2 Practice. It will also be useful for those candidates who are considering taking the Achievement Measurement Test 2 (AM2). Chapter 7, Revision Exercises, has been included to help candidates prepare for examinations such as the City and Guilds Course 2360 Part 2 Theory and other appropriate SCOTVEC and BTEC Courses.

Level 3 NVQ

Candidates who successfully complete assignments towards City and Guilds 2360 Theory and/or Practice Part 2 can apply this success towards Level 3 NVQ through a process of Accreditation of Prior Learning. There are additional practical tasks, to those required for City and Guilds Course 2360 Part 2 Practice, set in this studybook to assist those who wish to obtain their Level 3 NVQ.

Electrical Installation Series

Foundation Course

Starting Work
Procedures
Basic Science and Electronics

Supplementary title:
Practical Requirements and Exercises

Intermediate Course

The Importance of Quality
Stage 1 Design
Intermediate Science and Theory

Supplementary title:
Practical Tasks and Revision Exercises

Advanced Course

Advanced Science
Stage 2 Design
Electrical Machines
Lighting Systems
Supplying Installations

Study guide

This studybook has been written to enable you to study the underpinning background knowledge that will be required to complete the City and Guilds 2360 Electrical Installation Practice Part 2. It is likely that you will be asked to study the relevant material in advance of completing suitable practical tasks at a college, training centre or workplace under tutor supervision. The following points may help you.

- The practical tasks suggested in this studybook are for guidance only and your training centre/college may devise different ones according to local conditions, practices or preferences. Whether the exercises are the same or not they should all cover the same objectives and tutors will be looking for the same competences in the student. The practical work should be carried out under the supervision of a tutor in a suitable workshop. Guidance is given in this book to help you to carry out the practical work with greater understanding.
- Each task is preceded with some theory which should be studied before doing the practical work. The tasks have a check list entitle "Points to consider" which should assist you in being aware of what is expected of you. In addition you should be aware that you tutor will also be noting the following points:
  – Your ability to work with others (colleagues, supervisors and customers)
  – Your ability to treat visitors correctly
  – Your ability to contribute to your organisation's services to its customers including providing technical functional information to relevant people on handover
  – Your ability to adhere at all times to current regulations, recommendations and guidelines for health and safety
- A record of your achievements will be kept for you by the training centre/college and you will be able to have access to it in order that you are aware of your progress.
- It will be helpful to have available for reference a current copy of BS 7671:1992 and IEE Guidance Notes 3 "Inspection and Testing". At the time of writing BS7671 incorporates Amendment No.1, 1994 (AMD8536), Amendment No. 2, 1997 (AMD 9781) and Amendment No. 3, 2000 (AMD10983).
- Your safety is of paramount importance. You are expected to adhere at all times to current regulations, recommendations and guidelines for health and safety.

Your tutor may give you a programme of work. The boxes below may be used to assist you to complete your work on time.

Study times					
	a.m. from	to	p.m. from	to	Total
Monday					
Tuesday					
Wednesday					
Thursday					
Friday					
Saturday					
Sunday					

Programme	**Date to be achieved by**
Chapter 1	
Chapter 2	
Chapter 3	
Chapter 4	
Chapter 5	
Chapter 6	
Chapter 7	
Chapter 8 – Revision exercises – Section one	
Chapter 8 – Revision exercises – Section two	
Multichoice questions	

Practical tasks

Below is a complete list of all the practical tasks contained in this book. Tasks 1 to 45 have similar objectives to those in the City and Guilds Electrical Installation Practice Part 2. Tasks 46 and 47 are supplementary tasks to provide additional suitable situations that may be used to generate evidence to satisfy relevant performance objectives for NVQ Level 3.

Where suggested times are given in the tasks these are for guidance only.

Contents

6 Fault Diagnosis and Rectification 93

7 Revision Exercises 103

Multi-choice questions 111

Appendix 115

Answers 123

1

Preparation for Work

On completion of this chapter you should be able to:

- demonstrate the application to electrical installation work for building regulations, Health and Safety requirements, electrical regulations, British Standards and Codes of Practice
- prepare a site survey and carry out preliminary assessments for structure, fabric and installation factors
- prepare requisition lists from given site diagrams
- prepare specifications for:
  - installation materials
  - tools
  - equipment
  - and labour
- prepare a planned schedule from contract requirements and specifications
- prepare reports on "mock" accidents

Introduction

When undertaking the practical exercises required for this course you will be expected to demonstrate your ability to work with others. It is important that you can establish and maintain effective working relationships with not only your colleagues but also with customers and visitors. Your tutor will be taking your attitude into account during the completion of any of the practical exercises.

Also whilst completing the exercises you will be expected to take health and safety factors into consideration. You will be expected to adhere to current regulations, codes of practice and guidelines. These have been covered in the other books in this series but just to refresh your memory a summary is given here.

Regulations, Standards and Codes of Practice

Health and safety regulations

Keeping everyone safe at work is the responsibility of the employer and the employee and both are required, by law, to observe safe working practices. Various Acts of Parliament govern what employers provide in a workplace and how the employees use this provision.

Health and Safety at Work Etc. Act 1974

This Act applies to everyone at work and it sets out what is required of both employers and employees. The aim of this Act is to improve or maintain the standards of Health, Safety and Welfare of all those at work.

Under the Health & Safety at Work etc. Act 1974 are other Regulations and Codes of Practice including:

Management of Health & Safety at Work Regulations 1992
The Electricity at Work Regulations 1989
Manual Handling Operations Regulations 1992
CDM Regulations

There are other laws and regulations which deal with Health, Safety and Welfare at Work. Some of those with reference to the electrical industry include:

The Factories Act 1961
Safety Representatives and Safety Committees 1977
Notification of Accidents and General Occurrences Regulations 1980
Control of Substances Hazardous to Health Regulations 1988 (COSHH)

Strict observance of the applicable laws, standards and codes of practice is imperative to reduce the possibility of accidents. Another law that must be observed with particular regard to electricity is

The Electrical Supply Regulations 1988

BS 7671 Requirements for Electrical Installations, the Regulations published by the Institution of Electrical Engineers, are not law but are accepted as standard practice for electrical installation work. Guidance Notes issued by the Wiring Regulations Policy Committee of the Institution of Electrical Engineers can help as they give further details on the requirements of BS7671.

Guidance Notes available are:

Guidance Note 1 – Selection and Erection

Guidance Note 2 – Isolation and Switching

Guidance Note 3 – Inspection and Testing

Guidance Note 4 – Protection Against Fire

Guidance Note 5 – Protection Against Shock

Guidance Note 6 – Protection Against Overcurrent

Guidance Note 7 – Special Locations

The British Standards Institution (BSI) also produce Standards and Codes of Practice. Any which are applicable to the work activity should be complied with.

These Laws, Standards and Codes of Practice relate to the manufacture, installation and use of electrical equipment. Before working on electrical equipment all staff should be properly qualified, trained and competent in their work.

It is important to be able to find your way around Regulations, Codes of Practice and Guidelines. The following notes should help you to do this so that relevant information can be found quickly and accurately when required.

Table 1.1 shows a comparison between the main headings in Part II of the Electricity at Work Regulations 1989, BS7671 and CENELEC harmonisation documents (to 1991).

Table 1.1

Electricity at Work Regulations	BS7671 (IEE Wiring Regulations)	European Harmonisation Documents (CENELEC)
Systems, work activities and protective equipment	Part 1 and Part 3	
Strength and capability of electrical equipment	Part 1 and Part 5	HD 384.5.51
Adverse or hazardous environments	Part 1 and Part 6	To be published
Insulation, protection and placing of conductors	Part 1 and Part 5	
Earthing or other suitable precautions	Part 1, Part 4 and Part 5	HD 384.5.54
Integrity of reference conductors	Part 4 and Part 5	
Connections	Part 1, Part 5 and Part 6	
Means of protecting from excess of current	Part 1 and Part 4	HD 384.4.41, HD384.4.473
Means for cutting off the supply and for isolation	Part 1, Part 4 and Part 5	HD 384.4.46, HD 384.5.537
Precautions for work on equipment made dead	Part 1, Part 4 and Part 5	HD 384.4.46
Work on or near live conductors		
Working space, access and lighting	Part 1	
Persons to be competent to prevent danger and injury	Part 1 and Part 2	

BS7671 Numbering System

It is preferable to consider BS7671 in three parts:

- the regulations
- the appendices and
- the tables

The Regulations

The Regulations use a "three-part" numbering system. This is in line with the system used in the European harmonisation documents and that adopted by the European countries generally.

You will require to have available a copy of BS 7671 in order to work through the following exercise.

For example:

Regulation number 461-01-07

[4] 61 – 01– 07	Part 4 – Protection for Safety
[46] 1 – 01 – 07	Chapter 46 – Isolation and Switching
[461] – 01– 07	Section 461 – Isolation
[461 – 01] – 07	Group of Regulations
[461 – 01 – 07]	The seventh Regulation in that group

The Appendix

There are only six sets of appendices in BS7671 and these vary in size from three pages in Appendix 5 to 69 in Appendix 4. Although each appendix is given a number, the way the information is broken down depends on its complexity. If you turn to Appendix 4 note 6 you will see that two are subdivided again giving a three part number, 6.2.1. Note that dots are used between the numbers unlike the regulations which use dashes. Most of the appendices use tables to convey the information. Appendix 4 is the best example as this contains the current carrying capacity tables.

Tables

Tables found in the Regulations take their number from the Chapter they relate to. This means that the table number starts with two digits followed by a letter. For example Table 41B1(a) is found in Chapter 41 of Part 4.

[4] 1B1(a)	Part 4
[41] B1(a)	Chapter 41
[41B1] (a)	The second set of tables in chapter 41
[41B1(a)]	The actual table number

Tables found in the Appendices take their number from the Appendix number they are in. The table number in this case starts with a single digit followed by a letter and then a number and so on. For example Table 4D1A

[4] D1A	Appendix 4
[4D] 1A	Fourth group of tables in Appendix 4
[4D1] A	First table in this subgroup
[4D1A]	The actual table number

Try this

Look up the appropriate Regulation number in BS7671for the following:

1. Bends in conduit

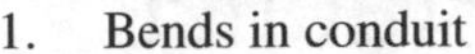

2. Labels - at the point of connection to any extraneous conductive part.

3. The position of a BSEN 60309-2 socket outlet in the vicinity of a swimming pool.

4. Use of rcd as protection against direct contact.

CDM Regulations

The Construction (Design and Management) Regulations 1994 (CDM Regulations)

It is one of the responsibilities placed upon the main contractor, that a Health and Safety plan is developed before construction starts. This plan is designed to ensure that proper provisions are in place for the duration of the contract.

Although the main contractor has overall responsibility for the administration of Health and Safety under these regulations, the co-operation of all sub-contractors is essential for compliance. The requirements and prohibitions placed on the contractor are listed under Regulation 19 of the CDM Regs.

It is a feature of most contracts which have been placed since the CDM Regs were adopted that the electrical contractor is required to provide all relevant information necessary for compliance in terms of planning, personnel and provision for health and safety. This will normally include method statements to provide details of how particular aspects of the project are to be carried out, along with an analysis of the risks to health and safety associated with the various tasks involved.

This information will be incorporated in a health and safety plan for the contract, which is distributed to all parties. It is the duty of all sub-contractors to support the principal contractor in fulfilling his duties under the regulations.

1 Regulations, British Standards and Codes of Practice

Objective: To demonstrate the application of regulations, British Standards and Codes of Practice to electrical installation work. (Amend according to resources available)

Suggested time: 3 hours

1. Identify which of these regulations:
   (a) Building
   (b) Health and Safety
   (c) BS7671 (IEE Wiring Regulations)
   (d) Electricity at Work (1989)
   (e) Electricity Supply Regulations (1988)

apply to **each** one of the following electrical installation practices:

(i) Testing the insulation resistance of a radial socket outlet circuit in a commercial building.
(ii) Drilling a hole through a floor joist in a three bedroomed detached house to accommodate the following twin and earth PVC/PVC cables: $1 \times 6\text{mm}^2$, $2 \times 2.5\ \text{mm}^2$ and $2 \times 1.5\ \text{mm}^2$.
(iii) Installing a 25 mm^2 C.N.E. service cable to a domestic property.
(iv) Wiring of emergency stop buttons in a machine shop which has six lathes.
(v) Fixing a sign on a main switchboard panel in a large factory to indicate voltage levels of 11 kV present.

2. (a) Identify the relevant British Standard(s) for **each** of the following:
   (i) shaver socket outlet installed in a bathroom
   (ii) 13 A switched connection unit installed in a kitchen
   (iii) 400 V 32A 3 P+N+E splashproof industrial cable coupler
   (iv) 100 A service fuse
   (v) S.B.C. lampholder

   (b) For **each** British Standard chosen identify a specific design feature in compliance with that standard.

3. BS7430 (1991) formerly C.P. 1013 (1965) covers the Code of Practice for earthing. State three requirements of this code of practice.

References:
Building Regulations
Health and Safety at Work Regulations
BS7671 (IEE Wiring Regulations)
Electricity at Work Regulations (1989)
Electricity Supply Regulations (1988)
Relevant British Standards

Points to be considered:

✓ have the correct regulations been identified?

✓ have the correct British Standards been identified?

✓ do the design features comply with the relevant British Standards?

✓ are the requirements stipulated applicable to the Code of Practice?

The site survey

During the course of the preparation for the design and compilation of the record data for the completed installation we may need to undertake some site surveys. These will involve measurement, marking out and recording details. A site survey may also involve access to areas of the site which may require security clearance, access arrangements and special access equipment.

Access

Before we can undertake a survey, we must obtain the co-operation of the site owner and the occupier in order that we can obtain access to the site. We need to confirm the details of the areas where we require access and provide some indication of the extent of the work we are intending to undertake during the survey. The user of the site will need to know the duration and the likely disruption that will be caused in order to agree a suitable time for the survey to take place.

Insurance and authorisation

In addition to the need to arrange and agree the physical access to the site we must also ensure that the appropriate assurances are in place. Contractors should have a public liability insurance to cover their everyday activities. In some instances we may need to obtain authorisation for the work we are to undertake. If we are to have access to "sensitive" areas with particular security requirements or risks, we are going to need suitable written authorisation to carry out a survey. This may be due to the need to gain access to, for example, HV switch rooms. In this case we would need not only the authorisation but also a permit to work.

The survey

Having considered the necessary arrangements for the survey to be undertaken, we need to examine the detail of the survey itself.

A survey may fall into a number of categories dependant on the information which needs to be obtained. The greater the extent of the information required the more involved both the survey, and the methods required to achieve the end result, will be.

The type and extent of a survey are dependant upon what it is intended to achieve and the information that is to be collected.

Visual survey

This survey, whilst beginning as a visual survey, often evolves to require some measurement or at least a counting exercise. A common example would be a survey to establish whether there are any adverse conditions involved along a proposed route for a new distribution cable (sub main) in an existing building. If no such conditions exist then there is probably no requirement other than the visual survey of the route. However if some adverse conditions are found, such as a high ambient temperature in a particular area, then measurement of the length of run and possible alternative routes would be necessary.

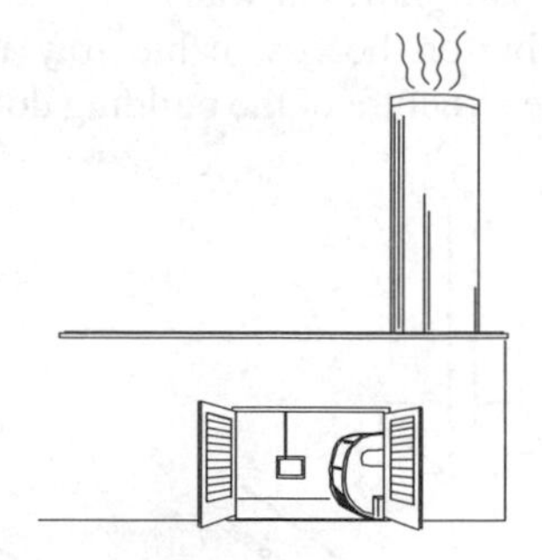

Figure 1.1 *A boiler room*

Site access survey

The most common and simple surveys we are likely to be involved in is to establish physical access to the site, or a suitable location for the set up of the storage and office facilities. We carry out such surveys, in their most basic form for every job we undertake.

Location, costing or manufacturing survey

Where detailed drawings are not available a more thorough survey may be required in order to establish requirements for quantities of materials for installation. Specialist installers may need to carry out detailed surveys with accurate measurements in order that bespoke equipment can be produced to meet the space confinement of the location, such as main control panels and the like. It is common for a detailed survey to be carried out to establish quantities of materials actually installed.

Structural survey

Wherever we are carrying out an installation in an existing building, some degree of structural survey will be required. The extent of the survey will depend on the nature of the building and the decorative finish. Without prior knowledge it may not be possible to tell from a visual inspection whether an internal wall is constructed of brick, blocks or timber or whether it is plastered or dry lined. We are usually obliged to carry out some survey, however small, to establish the nature of the building fabric before we commence work.

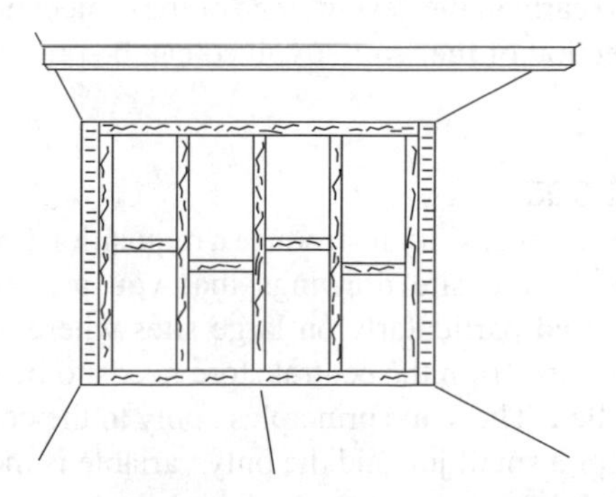

Figure 1.2 *Establish the nature of the building fabric*

In connection with the electrical installation it may be necessary to carry out a structural survey to establish whether areas are suitable for carrying the required loads. Lightweight fibre or plasterboard partition walls may not be suitable for mounting distribution boards, cable tray runs and heavy equipment. If the structure of the building does not lend itself

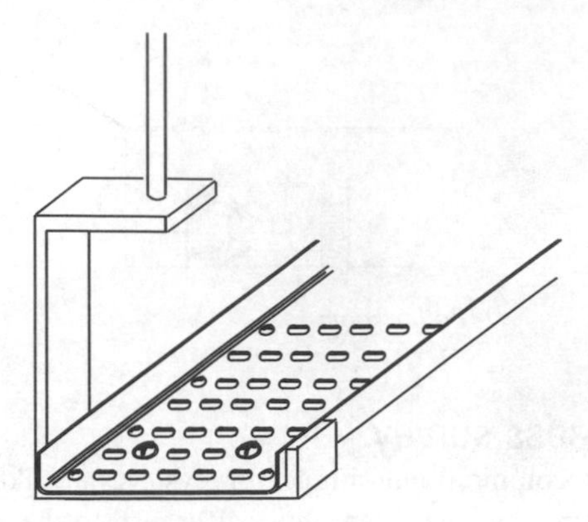

Figure 1.3 *Ceiling mounting may be necessary where wall mounting equipment is not appropriate*

to the proposed installation method then alternatives will need to be considered. This could involve a change to the type of installation or the use of alternative support methods. For example, we may need to install supports down from a concrete ceiling as opposed to wall mounting equipment.

It may also be necessary to carry out a structural survey in order to determine whether an existing building structure is adequate for the proposed additional work. We may be asked to install plant or equipment in a particular location within a building and it is necessary to ensure that the structure will support the weight of the equipment. This may be particularly significant with equipment such as generators, UPS equipment, chillers or air conditioning equipment and water storage vessels. It is usual in such cases for a structural engineer to be engaged to carry out the major structural surveys.

The extent of access or structural modifications or the need for particular installation techniques to be used can have a considerable effect on the cost of the work. Enabling and reinstatement work is essential but provides no added benefit for the client. Similarly a more costly installation technique as a result of the building structure results in no advantage or visible change for the client. These requirements need to highlighted early in the design stage of the project in order that a true reflection of the work involved can be presented.

Requisitions

It will often be necessary to produce a requisition for materials required from the site drawings that you are given. This process is used particularly on large sites where a record of materials issued from the central store needs to be monitored and controlled. The same principles apply to the collection of materials for a small job and the only variable is the extent of the work and the reference material used, drawing, job sheet or the like. We shall examine one method of doing this here but your company may use one of their own. It is important to make sure that, whatever method is employed, a logical approach is used when preparing the requisition for materials.

The important thing is to make sure that we do not miss or forget any items. To do this some logical method must be used to enable us to list, check and request all the material required. Remember that we will not necessarily be able to work uninterrupted on site and inevitably someone will always need something from us whilst we are preparing our requisition. It is therefore a good idea to develop a process which deals with the requirements in stages and where the progress made can be identified easily.

If we consider a general case such as a conduit installation then we will require:

First fix materials for conduit

First fix materials for wiring

Second fix materials for conduit

Second fix materials for wiring

Special accessories

So if we divide our requisition into these main stages then we can deal with each stage separately.

We may also subdivide each stage to give relatively small sections – for example:

First fix conduit:

Lighting
i) ground floor runs 1, 2, 3 etc.

Lighting
ii) first floor runs 1, 2, 3 etc.

Power
i) ground floor runs 1, 2, 3 etc.

and so on.

This provides us with a method of requisitioning materials that is in small stages so that work can be carried out efficiently and quickly and interruptions will not mean starting again from scratch.

There are special “take off” counters which operate rather like number stamps. Each time the device is depressed it increases the count by one and at the same time places a coloured mark on the drawing. This means that we can make a mark on each item, for example each twin socket outlet in an area, and keep a record of the number of items counted quickly, easily and reliably. These devices are particularly useful on large projects during the tender stage to establish quantity and hence cost. If such a device is not available then the use of coloured marker pens and a count notation system can produce similar results. A typical example of such a system would be coloured spot against each item with a coloured cross or number against every ten items, working up and down rows or working clockwise around the drawing. A good knowledge of the BSEN 60617 (BS 3939) symbols will be a definite advantage here.

However, it is common practice amongst architects and consulting engineers to use "non-standard symbols", in which case, a key to the drawing symbols is essential.

Figure 1.4

Remember

When making a survey consideration must be given to:

Environment

safety
type
dimensions
access (for example, for delivery of materials)
security
access equipment (any special equipment required?)
other tradespeople on site

Building structures

safety
type
materials
dimensions
access
security

Existing services

electrical wiring	safety, type and location
	supply position and capabilities
gas supply equipment	type and location
water	type and location
alarm systems	type and location
heating systems	type and location
ventilation	
natural light	

Other factors relative to new wiring

installation materials	
tools	
equipment	
labour required	
customer's requirements	including preferences
	cost
other tradespeople	

This list is not exhaustive. Please add any other factors you feel are important.

2 Prepare a site survey

Objective: To prepare a site survey and carry out preliminary assessments for:
(a) structure, (b) fabrics, (c) installation factors

Using the specifications and drawings for the project (CT Manufacturing Ltd) in "Stage 1 Design" book (Refer to Appendix for details and specifications) or an alternative as specified by your tutor:

(a) Prepare a site survey
(b) Carry out preliminary assessments for the building structure and fabric, and electrical installation factors.

CT Manufacturing Ltd

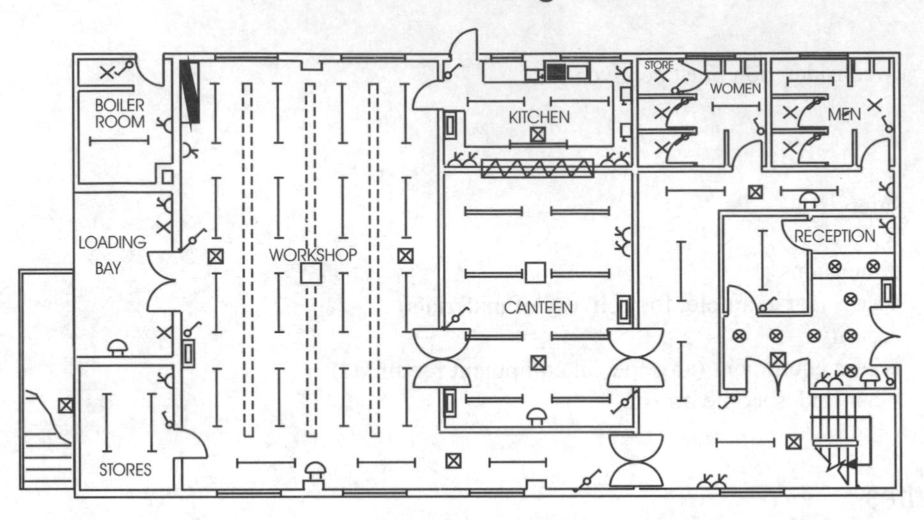

GROUND FLOOR PLAN

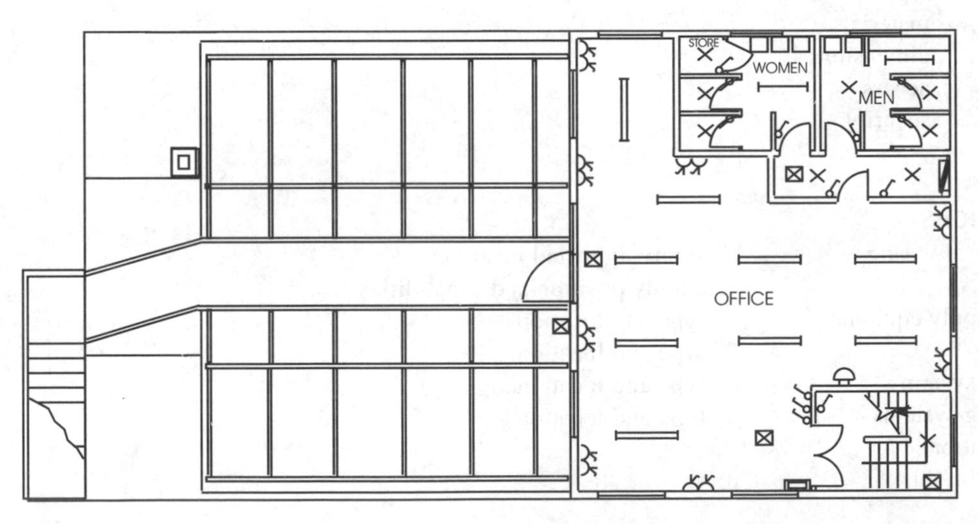

FIRST FLOOR PLAN

Figure 1.5

Time allowed: 2 hours

Have the following points been taken into consideration?

	Tick box Yes	No
Customer's requirements	☐	☐
Health and Safety factors	☐	☐
External influences	☐	☐
Environmental conditions	☐	☐
Type of building	☐	☐
Building dimensions	☐	☐
Building structure	☐	☐
Building fabric	☐	☐
Electrical wiring	☐	☐
Electrical power points	☐	☐
Lighting	☐	☐
Alarm equipment	☐	☐
Heating and ventilation	☐	☐
Other services	☐	☐
Compliance with EAW Act and BS7671	☐	☐

3 Requisitions

Objective: To prepare requisition lists from site diagrams and specifications.

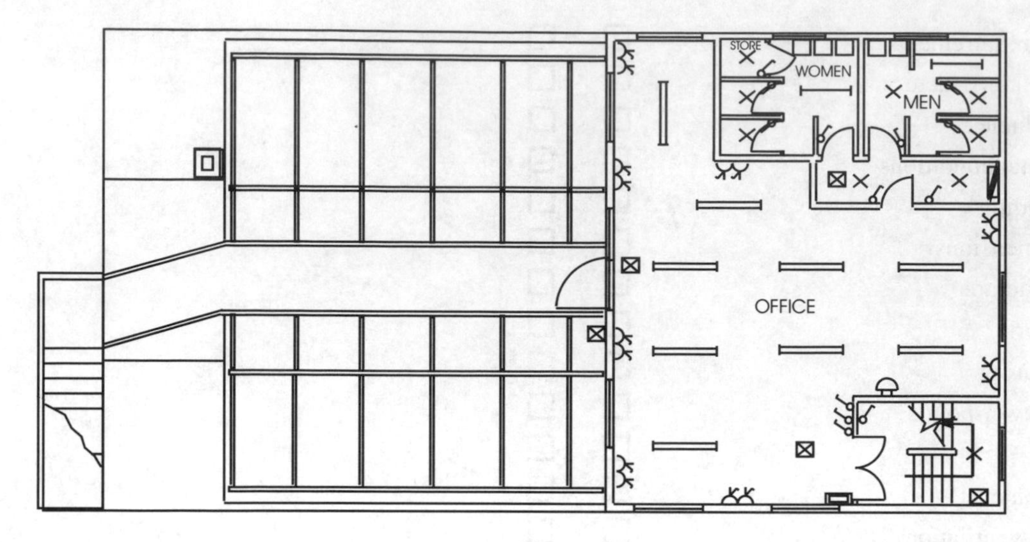

Figure 1.6 *Scale 1:200*

Suggested time: 2 hours

Method:

1. Using the drawing in Figure 1.6 make a requisition for the second fixing materials required to complete the work.

2. Using the specification on p.116 compile a list of any special tools and access equipment required for the work involved.

3. What labour will be required to first and second fix this installation?

Points to be considered:

✓ are all materials required listed?

✓ are all special tools required listed?

✓ has suitable access equipment been chosen?

✓ are the labour requirements suitable?

Planned schedule

Where work involves a number of trades there are likely to be several subcontracts awarded. Each subcontractor will need to develop a planned schedule for the completion of their work. Each schedule will need to be co-ordinated with the others to ensure that the work is carried out within the time allowed and with the minimum of disruption to progress.

In order to achieve this it is common for the contractor responsible for the construction to produce an outline programme for the construction aspects of the work. Each contractor will produce a programme of their own aspects of the work. These will then have to be put together to form a working programme for the project.

The extent to which the individual programmes can be accommodated will depend on the overall programme time and the time available for each stage of the work. Compromise is almost always necessary with contractors working together to produce a workable programme. Because of this it is common to find trades sharing work areas and tight schedules. If one aspect of the work is delayed, it can have serious repercussions for the activities which follow and it is for this reason that delays can result in considerable additional costs to the project.

The basis for the original programme is quite simple, the extent of construction or alteration to an existing structure is of prime importance as until a building exists there can be no work carried out within it. Likewise the building does not have to be, and generally cannot be, complete before work can commence on the services and internal construction. The approach needs to be logical and well thought out, for each activity there is some necessary preliminary work which must be in place before the activity can begin.

We must also have the necessary material, switch box, wall plugs and screws all of which are fairly obvious. But we must also ensure that we can get to the location and there are not obstructions such as scaffolding, shuttering, plant or materials in the way. Equally important we must ensure that there is not other work scheduled in the area, such as laying flooring, spraying, overhead construction, welding and the like, which would prevent us from carrying out our work.

On a large project there are many aspects to be considered and in order to achieve the project completion on time the works must be integrated as effectively and efficiently as possible. Each contractor has an obligation to complete their activity within the programmed time and to the programme dates.

Planning

For a small installation we need to consider obtaining material, arranging access, and planning when our next job is due to start.

We have considered all the activities necessary to produce a basic programme for our work using a simple logic sequence for carrying out the installation work.

Once the order is received we need to:

- order material and organise labour for the start date
- deliver 1st fix material and ensure the labour is on site for the start date
- deliver 2nd fix material to the site just in advance of the second fix
- inspect and test the installation
- complete the documentation and certification for the work

During the site activity we need to monitor the progress of the work to ensure that any deadline dates are achieved. These are usually to allow "follow on" trades, such as plasterers, carpenters and decorators to complete their work on time.

This is an oversimplified procedure but the intention is to provide us with an idea of the approach required when we consider more complex projects. This type of planning is generally relayed to a bar chart in order that the project can be viewed as a whole and the stages of construction and completion checked on a regular basis. To keep this simple each activity must be complete before the next activity begins.

See Chapter 7 in "The Importance of Quality" for more details.

Programmes

As construction projects become increasingly more complex, the planning and the execution to ensure completion at a given time becomes more important. One way in which this is achieved is by the production of Critical Path Networks. The principal function of these networks is to identify the critical elements in the project which must be completed at a given time to ensure the overall project is completed to programme. These items may appear to be unrelated in terms of trade or process but form a vital link in the completion of the project.

Figure 1.7

4 Prepare a planned schedule

Objective: To prepare a planned schedule from contract requirements and specifications.

Prepare a planned schedule for the electrical installation work at CT Manufacturing Ltd.

CT Manufacturing Ltd

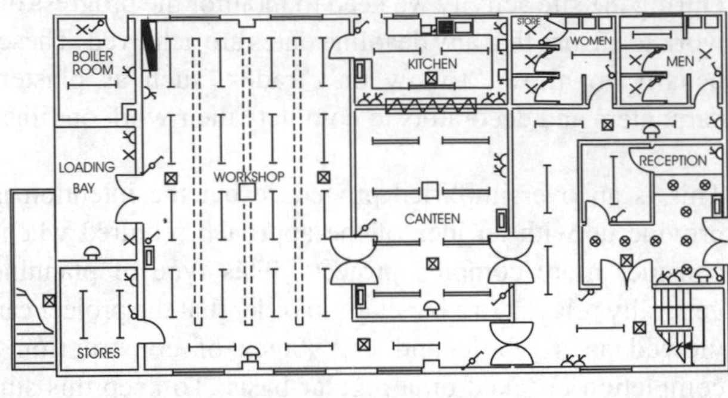

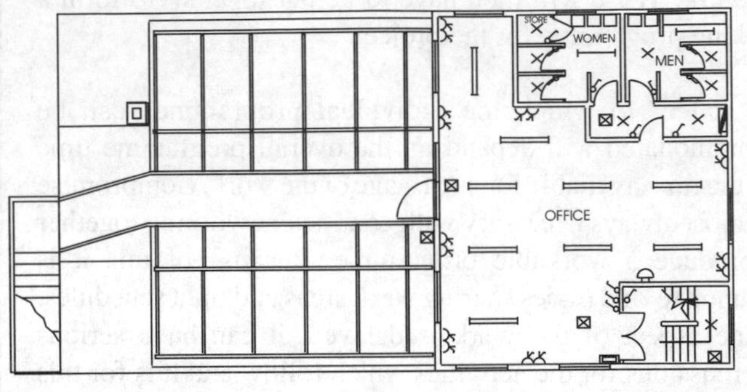

FIRST FLOOR PLAN

Figure 1.8

(Refer to Appendix for details and specifications)

Time allowed: 2 hours

Have the following points been taken into consideration?

	Tick box Yes	**No**
Start/finish dates	☐	☐
Labour requirements	☐	☐
Materials deliveries (1st and 2nd fix)	☐	☐
Contract requirements	☐	☐
All specification details	☐	☐
Installation stages	☐	☐
Co-ordination with other trades	☐	☐
Inspection and testing	☐	☐
Certification	☐	☐
Connection of supply	☐	☐

Accidents at work

Keeping safe at work is the responsibility of both the employer and the employee. A safe system of work should be developed which identifies hazards found in the workplace. The most frequent causes of accidents are falls, falling materials, electrical accidents, the lifting of heavy or awkward loads and mobile plant.

Ill health may be suffered as a result of exposure to hazardous substances or dust. High noise levels can cause partial deafness and vibration can cause "dead fingers".

Reporting accidents and work related diseases

The Reporting of Injuries, Diseases and Dangerous Occurrences Regulations 1995 (RIDDOR) require that certain accidents that happen on site have to be reported to the Health and Safety Executive.

This requires the notification without delay, usually by telephone, of serious and fatal accidents. This action must be followed by the presentation to the HSE of a completed accident form (F2508) within ten days. A similar requirement exists for dangerous occurrences such as the failure of a crane or lifting device or contact with overhead lines and collapse of a building or scaffold. Other less serious incidents must be notified to the HSE using the F2508 form within ten days of the incident.

Incidents of specific work related diseases must also be notified to the HSE by the use of an F2508A form.

Accident book

Wherever people are employed under the provision of the Factories Act, The Offices, Shops and Railway Premises Act or where ten or more people are normally employed then the employer is responsible for keeping an accident book.

Every accident at work, even the minor ones, should be recorded in the book and there are certain facts that must be entered.

An accident form should contain the following information:

- date and time of the accident
- place where the accident took place
- brief description of the circumstances
- name of the person injured
- sex of the person injured
- age of the person injured
- occupation of the person injured
- nature of the injury sustained

Serious injuries are referred to as major injuries and these are defined as:

- fracture of the skull, spine or pelvis
- fracture of any bone: in the arm, other than the wrist or hand, in the leg, other than in the ankle or foot
- amputation of a hand or foot
- loss of sight of any eye
- any other injury which results in the injured person being admitted to hospital as an inpatient for more than 24 hours

In addition every accident to an employee which results in the inability to work for more than three consecutive days must also be reported.

E.E.ELECTRICS LTD

Accident Report Form

This form must be completed in the event of any accident/injury occurring at the above premises or whilst working for the above company.

Date and time of accident:

Date: *Time:*

Place where accident took place:

Brief description of the circumstances:

Injured person:

Name: *Sex:* *Age:* *Job title:*

Nature of injury sustained:

Your details:

Name: *Job title:*

Figure 1.9

5 Accident Report Forms

Objective: To prepare a report on a "mock" accident.

Suggested time: 1 hour

Method:

Prepare and "fill in" an accident report form using the following information.

Colin Conduit of 6 Edison Avenue, Burton (aged 26) is an electrician working for E.E.Electrics Ltd of Faraday Road, Derby which employs a total of 22 JIB graded electricians.

At 11 a.m. on 1st April 1999 he was working at J.Joules (Jewellers) shop, Henry Street, Derby with subcontractor T.Trunking.

He was working 3 metres up a ladder over the shop window drilling the wall with a portable electric drill. He drilled into a live buried electric cable, received an electric shock and flash burns and fell from the ladder to the pavement fracturing his skull. An ambulance was called and he was taken to the local hospital. As a result he was off work for 3 months.

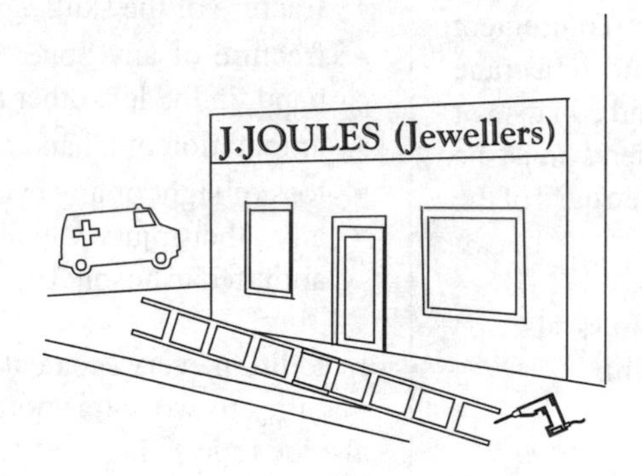

Figure 1.10

Points to be considered:

✓ does the form comply with the Health and Safety at Work Act?

✓ is the form documented correctly

✓ is the information given on the form correct?

2

Isolation Procedures and Measuring

On completion of this chapter you should be able to:

- demonstrate how to safely isolate and carry out relevant confirmatory test(s)
- use appropriate instruments to measure current on single phase and three phase four wire systems
- use appropriate instruments to measure voltages on single phase and three phase four wire systems
- use appropriate instruments to measure current and voltage drop in a radial circuit with several load positions
- compare readings obtained from wattmeters with the calculated value of circuit wattage based upon the readings of voltmeters and ammeters.

The tasks in this chapter can be carried out using the simulated distribution system shown on p.17 or another similar system that is provided by your college or training centre.

Supply systems

A supply system can cover the supply of electricity from the substation at 11 kV down to the consumer and throughout the installation. It can also apply to the supply from sources such as batteries and power supply units.

It is quite common to find a factory with a substation transformer adjacent to it, or even built into the premises. The supply to the transformer is usually 11 kV to a delta connected winding. This means that only three conductors are connected to this side of the transformer. The output is normally from a star connected winding delivering the voltages required within the factory. These voltages can be categorised under two headings, namely single-phase and three-phase.

Single-phase

Single-phase supplies are usually associated with one phase and the neutral star point of the transformer windings. However, this is not always the case for single-phase supplies can be obtained by connecting to any of the two phases. The voltages are of course different in each case. The supply from any one phase and neutral will normally be 230 V, as shown in Figure 2.1, whereas connections between any two phases will give 400 V (Figure 2.2).

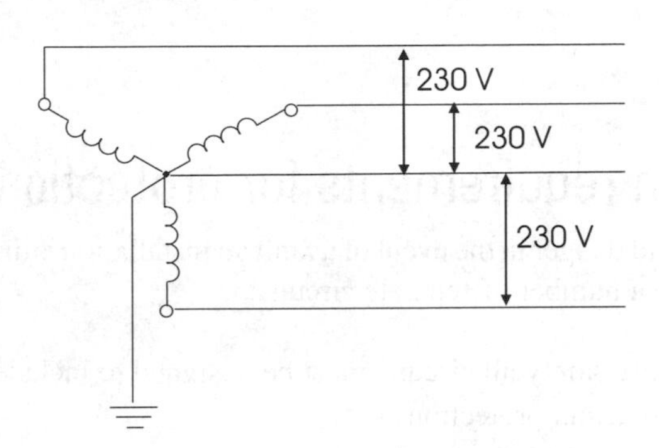

Figure 2.1

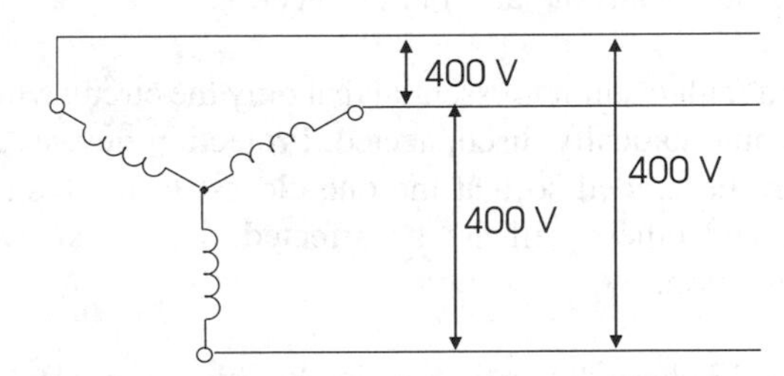

Figure 2.2

Three-phase

A three-phase supply is obtained, as the name implies, from all three phases. this normally gives a voltage of 400 V between any of the phases (Figure 2.3).

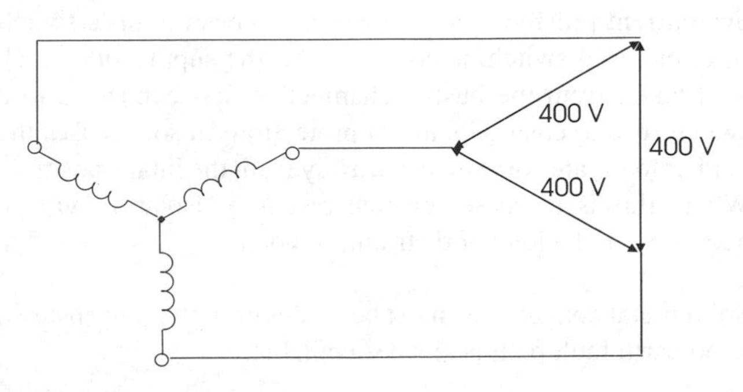

Figure 2.3

So where the output winding of a transformer is connected as a star with the centre point common to all three-phases, a combination of voltages can be obtained. These consist of 230 V and 400 V single-phase and 400 V three-phase.

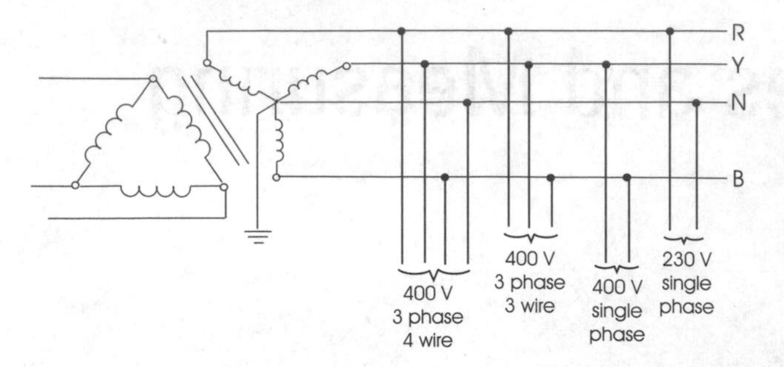

Figure 2.4

Main requirements for protection

To avoid danger in the event of a fault an installation is divided up into a number of separate circuits.

To ensure safety all circuits must be designed to include:

- overcurrent protection
- means of isolation
- adequate means of earth fault protection

Should a fault occur it is essential that only the circuit with the fault is automatically disconnected. Protection devices must therefore be graded so that the one closest to the fault will operate and others will not be affected. This is known as discrimination.

On a single-phase domestic type intake this is comparatively simple for each circuit has its own protection device within a consumer unit. The means of isolation is the main switch and the earth fault protection is either through the protection devices or an RCD which can be separate or can be used as the main switch.

Where a number of loads have to be taken to separate circuits at the main intake, such as on a three-phase supply, a busbar chamber is often used as a large junction box. This chamber, as with anywhere else on the installation, must have means of overcurrent protection and isolation. To ensure this a switch fuse, or fused switch, is connected on the supply side. Each load taken from the busbar chamber is first connected to a switch fuse to ensure its initial protection. In some cases the load or loads are some distance away from the intake position. Where this is the case separate means of isolation will be necessary at the load or distribution board.

Solid metal connections must be made throughout to ensure a good earth fault path is always available.

Before working on equipment check that it has been isolated and that it is dead. Following a procedure such as that in Figure 2.5 helps to ensure everything is checked.

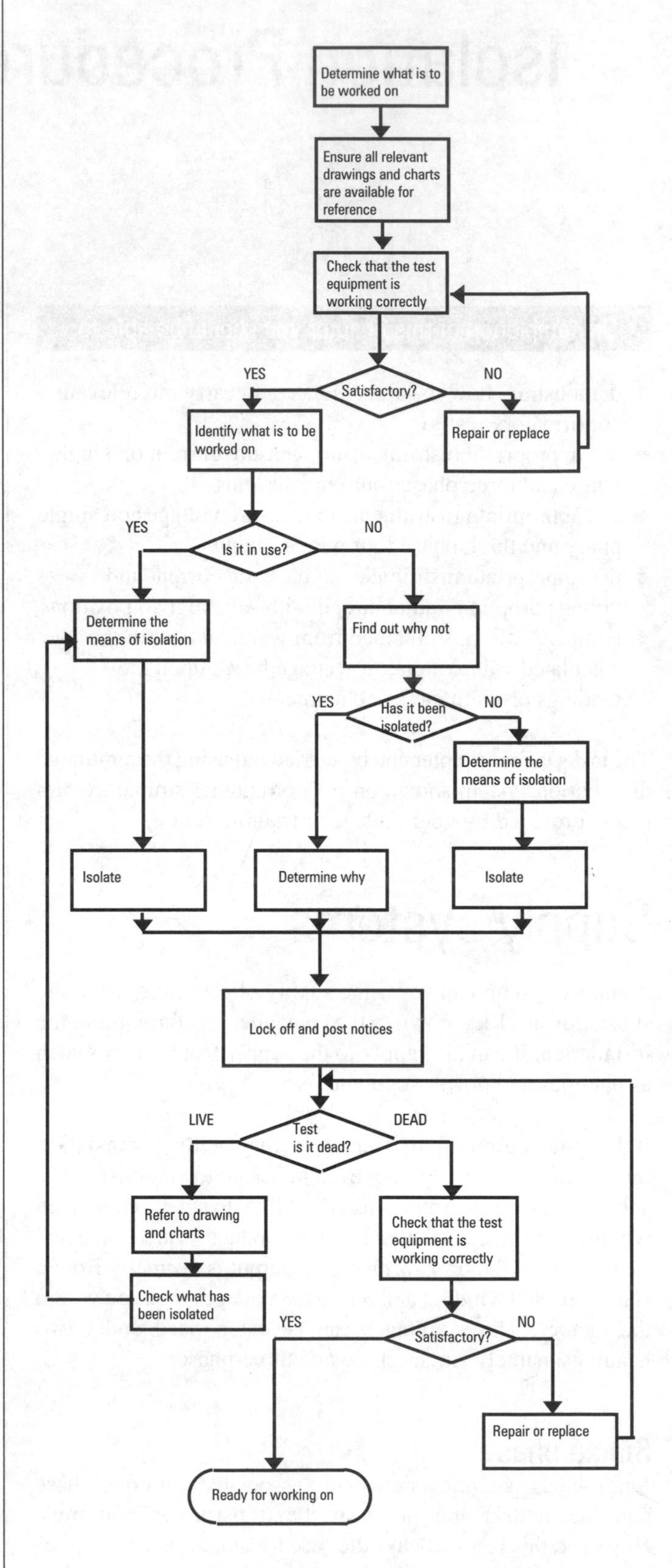

Figure 2.5 Isolation flow chart

Simulated Distribution System

This simulator consists of three 13 A socket outlets each connected to a separate phase using 2.5 mm^2 single PVC insulated and sheathed cable. The circuits all pass through a connector strip which is used for taking voltage readings.

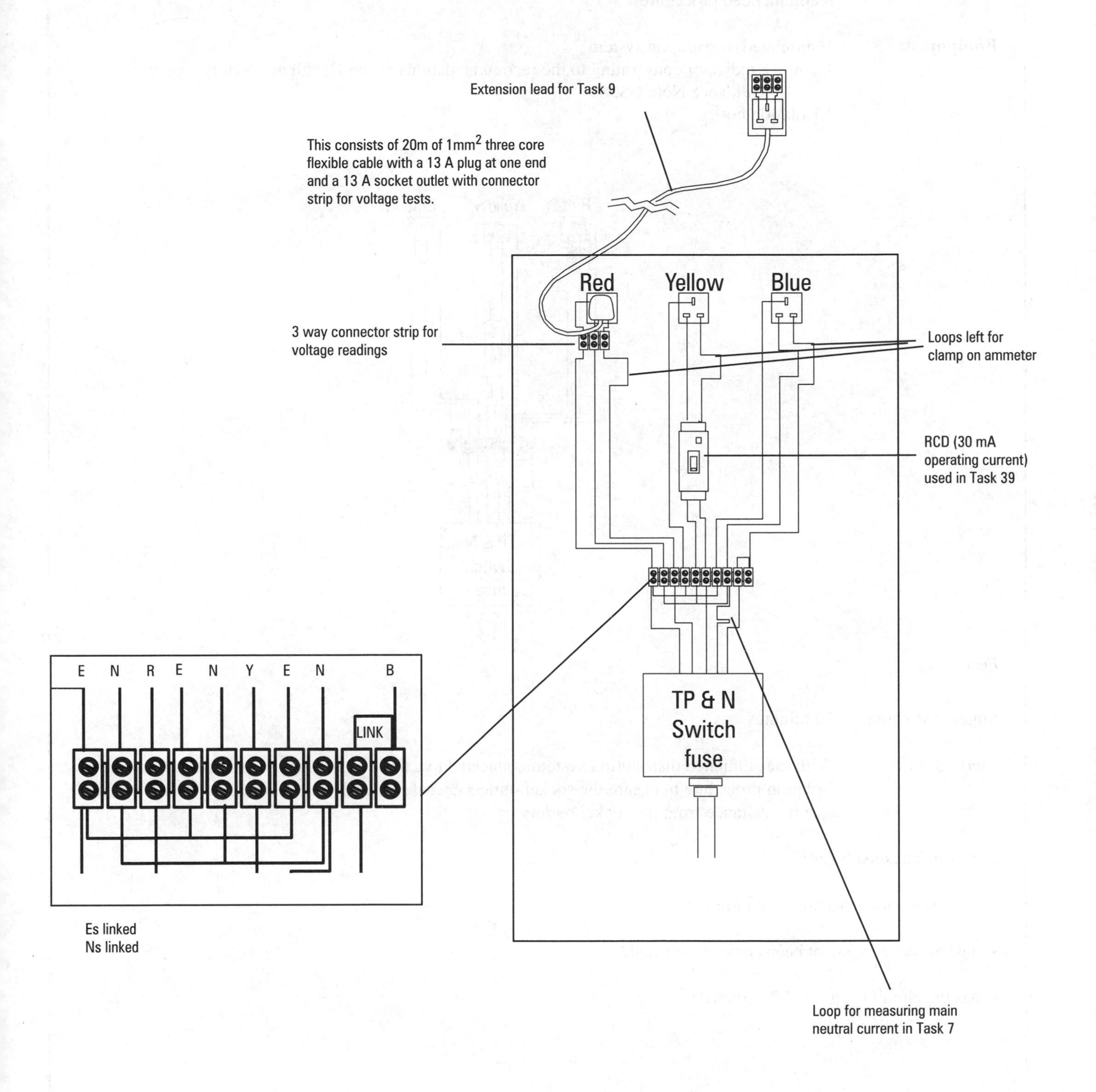

Figure 2.6 Simulated distribution system for Tasks 6, 7, 8, 9, 11 and 39.

6 Isolation procedures

Objective: To demonstrate the safe isolation of a section of an installation to recommended procedures.

Equipment:
1 simulated distribution system
1 voltage indicator conforming to the recommendations of the Health and Safety Executive Guidance Note GS38
1 isolation notice

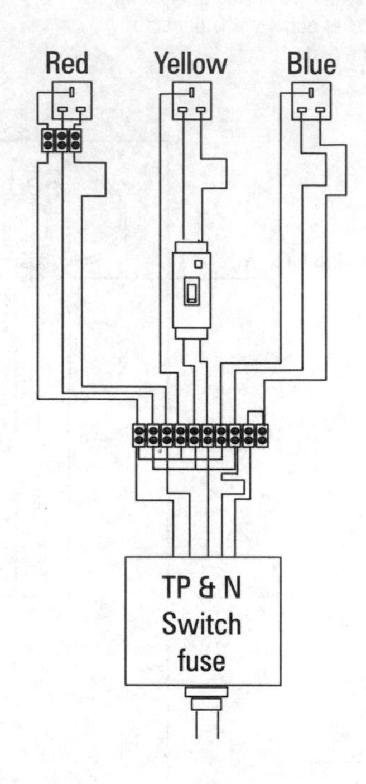

Figure 2.7

Suggested time: 30 minutes

Method: With the simulated distribution system connected to a three phase supply carry out the standard isolation procedure to ensure the socket outlets are safe to work on. Assume that the switch fuse is some distance from the socket outlets

Points to be considered:

✓ has a standard procedure been followed?

✓ has the test equipment been correctly verified?

✓ has the circuit been suitably isolated?

WARNING!
This exercise should be carried out in accordance with the Electricity at Work Regulations 1989 and all live conductors should be shielded and insulated from touch. All test probes should meet the requirements of the Health and Safety Executive Guidance Note GS38.

Monitoring and metering

It is not enough to know that a circuit works, it is also necessary to be able to check that it is working in the way that it was designed. It is often necessary to take the measurements of current, voltage and power to check that the equipment is working to its specification.

Current measurement

As current flows through a circuit to measure it the ammeter (flow meter) must be connected into the circuit. This means that the circuit has to be broken and the meter connected into it.

This can be dangerous as the connection leads and the meter have all got to be capable of carrying the maximum currents. These should never be connected when the circuit is energised.

Where current is to be monitored continually with panel mounted meters these meters may be wired into the circuit. It is more likely, especially if the currents are high, that current transformers would be used. A current transformer uses the magnetic field set up around a conductor carrying alternating current. The conductor acts as the primary and a secondary coil is placed around it.

This method of monitoring current means that the current carrying conductor does not have to be broken as it passes through the current transformer coil. Only a single conductor should be used because if a twin cable with a phase and return was put through the coil the ammeter would read zero. This is because the magnetic field set up by the phase conductor would be cancelled out by the opposing field in the neutral conductor.

As current transformers are basically double wound transformers, but in this case the primary winding has only one turn, the same calculations can be used. Assuming a conductor carries a maximum load of 500 A and the ammeter is calibrated to show this when in fact only 5 A is flowing through the meter, the coil must have 100 turns.

Clamp-on ammeter (tong tester)

The current transformer, which is fixed as shown in Figure 2.8, consists of a complete coil which has the load conductor passing through it. To change this from one conductor to another means the loads and supplies must be switched off, the coil disconnected and then reconnected on the new load and then it can all be switched back on again. This takes time and can be very inconvenient.

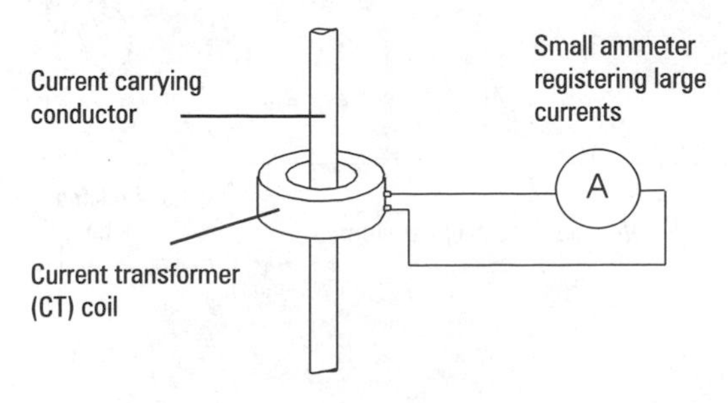

Figure 2.8

However, the theory using the current transformer has been adapted so that a portable easy to operate meter has been developed.

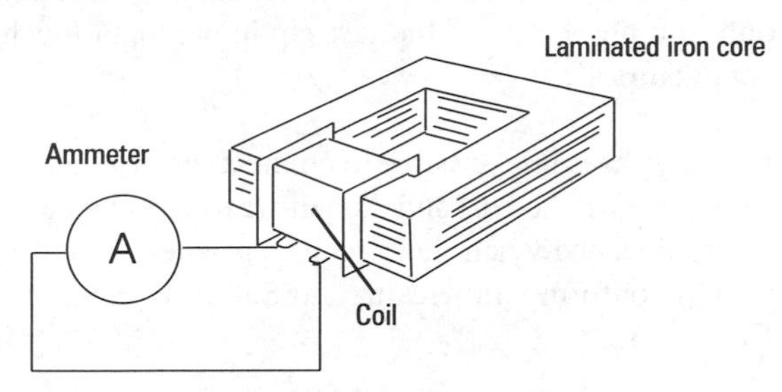

Figure 2.9

As with most double wound transformers the coil is wound on a former and fitted on a laminated core. This iron core is different to others insomuch as it is made to open up and allow room for a conductor to go into the centre.

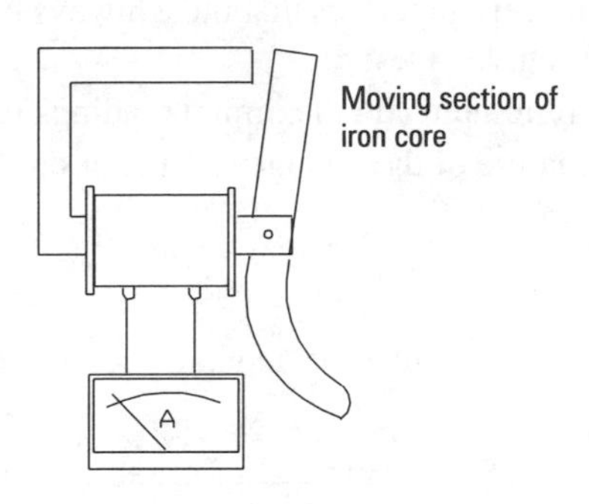

Figure 2.10

The moving section of laminated iron core is sprung so that when the gap is closed there is a tight joint between the two surfaces. The method of opening the core and where it opens depends on particular manufacturers. An example of a complete "clamp on" ammeter is shown in Figure 2.11.

This same method for monitoring the current can be used when measuring power and power factor. Some of the meters using this method have automatic range switching, others have an external switch. It is not uncommon for this type of meter to be capable of measuring up to 1000 A.

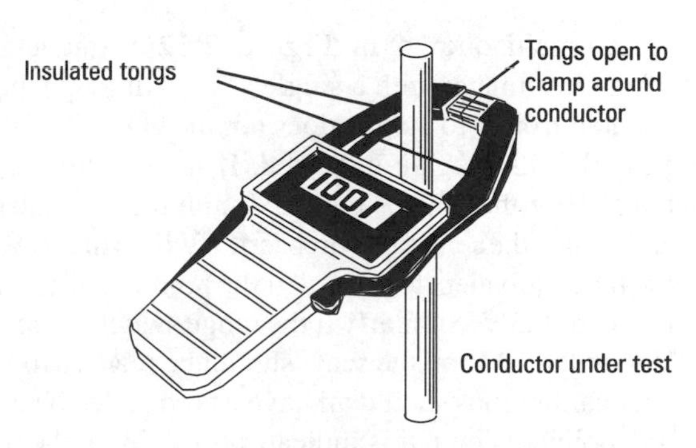

Figure 2.11 *Digital clamp on ammeter*

Voltage measurements

The fact that voltage readings must always be taken on live circuits means that every safety precaution should be taken. Barriers and screens may be necessary to offer protection from touching exposed live conductors. Wherever possible voltages should be taken where all live conductors are insulated so that only the probe tips of the test equipment can touch the live conductors.

In many cases it is possible to connect the voltmeter when the supply is switched off and then the readings can be taken from a safe distance when the circuit is re-energised. Test probes should conform with Health and Safety Executive Guidance Note GS38.

Voltmeters come in many different forms. They can be meters designed for one voltage range, say 0–200 V. In this case they cannot be used for voltages above that and voltages below 50 V would be inaccurate. Some of the digital meters are completely automatic and will select the range most suitable for the voltage being measured. Regardless of whether it is a single range or multi-range instrument, analogue or digital, there are certain factors that must always be considered when measuring voltages:

- the minimum and maximum readings required
- the nature of the voltage i.e. a.c. or d.c.
- safety

Let us now consider the actual use of an instrument.

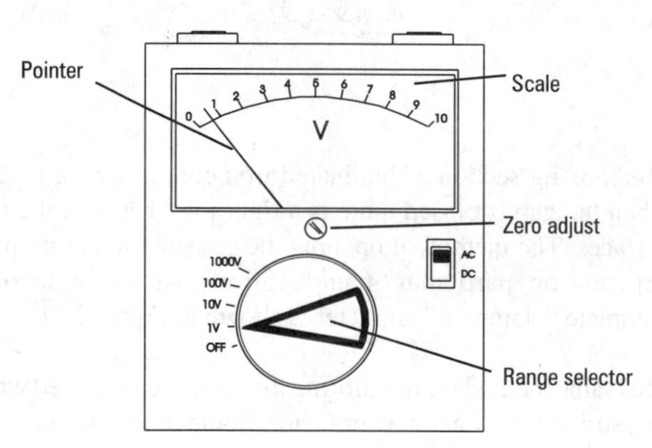

Figure 2.12 Voltmeter

The instrument shown in Figure 2.12 is an analogue multi-range voltmeter with a single scale. Although the scale is numbered from 0 to 10 this does not mean it will always be the 0 to 10 volts that are indicated. If the selector switch is pointed at 10 V then that is the maximum the instrument will indicate. When the selector switch is at 1 V then the 10 V on the scale will be equivalent to 1 V and if the pointer indicates 5 this will in fact be 0.5 V. Similarly if the range switch is pointing at 100 V then the 10 V on the scale should be read as 100 V and the other calibrations should all have a zero added on to them, so if the pointer is on 6 this indicates 60 V. With the selector switch on 1000 V then 2 zeros are added.

Voltage readings should always be taken across a difference in potential. For example across the supply.

Although a supply voltage, phase to neutral, is 230 V if a star connected three phase supply is used the maximum voltage available is 400 V. This means that if there is any doubt as to what the voltage may be the meter must be switched to read the maximum and then switched back if necessary.

The supply company will state what their nominal voltage will be at the supply intake. They have a legal obligation to keep that within + 10%, –6% of what they state. Within the building any drop in voltage due to long cable runs and large loads is up to the consumer and their consultants. Generally the voltage drop within an installation is kept to no more than 4% of the supply voltage. To ensure this has not been exceeded voltage readings sometimes have to be taken at the load and these compared with the supply voltage. Excessive voltage drop can cause the malfunction of equipment and heat being produced in cables.

Note: Domestic, commercial and industrial voltage values were changed from 415/240 Volts, ± 6%, to 400/230 Volts +10%, –6% on 1st January 1995. Voltage tolerance levels will be further adjusted to ± 10% by the year 2003.

Measuring high voltages

It is sometimes required to measure voltages that are up in the thousands of volts. An example of this is on the supply transmission and distribution systems. So that voltages can be monitored all the time voltmeters are built into control panels. To reduce the risk of danger and keep the voltmeter a practical size transformers are used on the high voltage cables.

Using this method the maximum voltage on the meter may be only about 50 V even though the scale is calibrated to read 11000 V.

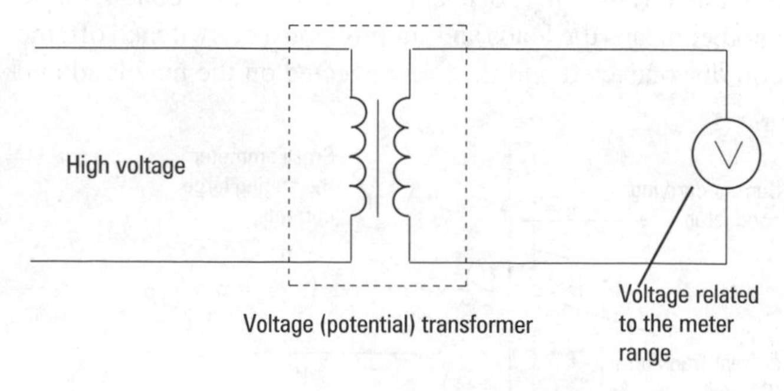

Figure 2.13

7 Measuring current on single and three phase circuits

Objective: To measure current using clamp on type instruments for three phase and single phase supplies.

Equipment

1 simulated distribution system
1 clamp on ammeter
1 wire-in ammeter 0–10 A a.c.
3 single phase loads approximately 1 kW each
1 load of 2 kW

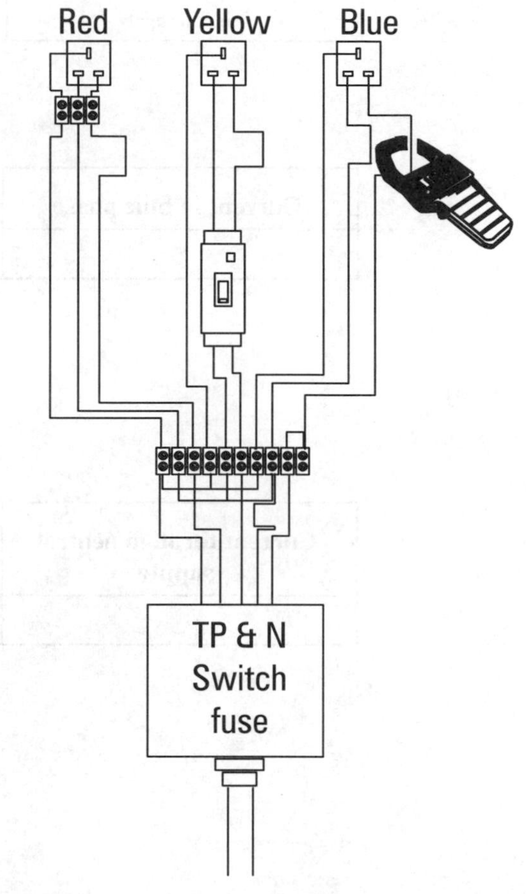

Suggested time: 1 hour

Figure 2.14

Method:

1. A three phase and neutral supply with earth leakage protection should be connected to the TP and N switch fuse.
2. Connect the three single phase loads to the socket outlets.
3. Switch on the switch fuse, take readings using the clamp on ammeter and complete the results table.
4. With the switch fuse in the OFF position remove the LINK on the Blue phase and connect the 0–10 A ammeter in its place. Switch on and note the readings on the meter.
5. Switch off the switch fuse, remove the ammeter and replace the link. Remove the load from the Red phase, replace with the 2 kW load, put the switch fuse on and using the clamp on ammeter measure the current in the neutral to the switch fuse.

WARNING!

This exercise should be carried out in accordance with the Electricity at Work Regulations 1989 and all live conductors should be shielded and insulated from touch. All test probes should meet the requirements of the Health and Safety Executive Guidance Note GS38.

Results:

For method 3

Current through conductor	Current (A)
Red phase	
Yellow phase	
Blue phase	
Neutral and blue phase together	
Neutral supply	

For method 4

Current in blue phase	Current

For method 5

Current through neutral supply	Current

Questions:

1. Explain the advantage of using clamp on ammeters over the wire-in type.

2. With the three loads connected in method 3 draw a phasor diagram and prove the neutral current.

8 Measuring voltage on single and three phase circuits

Objective: To measure voltages on three phase and single phase supplies.

Equipment: 1 simulated distribution system (Figure 2.15)
1 voltage measuring equipment

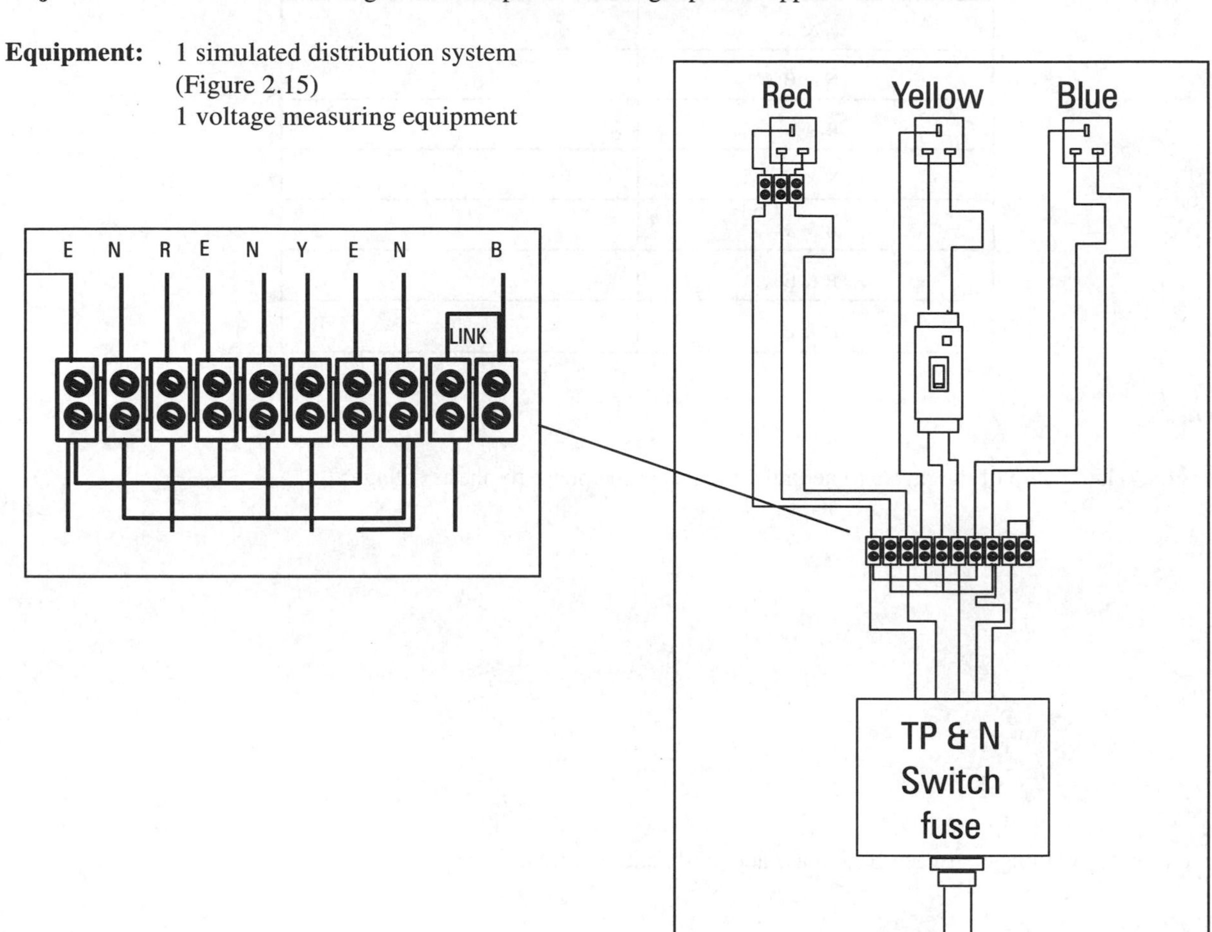

Figure 2.15

Suggested time: 30 minutes

Method:

1. A three phase and neutral supply with earth leakage protection should be connected to the TP and N switch fuse.

2. Switch on the switch fuse, take the readings and complete the results table.

Note:
If the voltage measuring equipment does not give the readings switch off the switch fuse and have the equipment checked.

WARNING!
This exercise should be carried out in accordance with the Electricity at Work Regulations 1989 and all live conductors should be shielded and insulated from touch. All test probes should meet the requirements of the Health and Safety Executive Guidance Note GS38.

Results:

Readings between	Voltage
N & R	
N & Y	
N & B	
R & Y	
Y & B	
B & R	
R & E	
N & E	

Questions:

1. What is the relationship of the phase to neutral voltages, to the phase to phase voltages?

2. Comment on the readings between Red phase and earth, and neutral and earth.

Red to earth:

Earth to neutral:

9 Measuring voltage drops on single phase circuits

Objective: To compare the voltage at a load position with that of the supply.

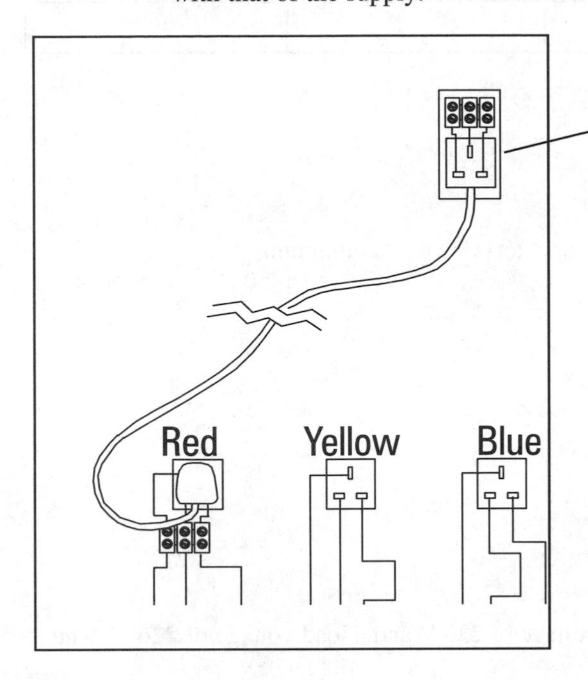

Figure 2.16

Equipment:
1 simulated distribution system
1 extension lead as shown
1 voltage measuring equipment
1 load of 1 kW
1 load of 2 kW

Suggested time: 30 minutes

Method:

1. A three phase and neutral supply with earth leakage protection should be connected to the TP and N switch fuse. The fuses for Yellow and Blue phases can be removed.

2. With the switch fuse in the ON position and the extension lead plugged in to the Red phase without a load connected, measure the voltage at the supply block and the load block and note the readings.

3*. With the 1 kW load plugged into the extension lead measure and note the voltage at the supply and load blocks.

4*. Replace the 1 kW load with a 2 kW load and repeat the voltage readings noting the results.

*Note: Mains voltage can fluctuate between readings. Reliable results can only be obtained by using 2 identically calibrated instruments simultaneously. This also applies to Task 10 on p.27.

WARNING!

This exercise should be carried out in accordance with the Electricity at Work Regulations 1989 and all live conductors should be shielded and insulated from touch. All test probes should meet the requirements of the Health and Safety Executive Guidance Note GS38.

Results:

	Voltage at supply	Voltage at load	Voltage drop at load
No load			
1 kW load			
2 kW load			

Questions:

1. Explain why it is important to keep the voltage drop on conductors down to a minimum.

2. If a current of 4.8 A is flowing in a cable with a supply voltage of 230 V and a load voltage of 226 V, what power is dissipated in the conductors?

10 Voltage drop in a radial circuit

Objective: To measure the voltage and current at different parts of a loaded radial circuit and compare the results.

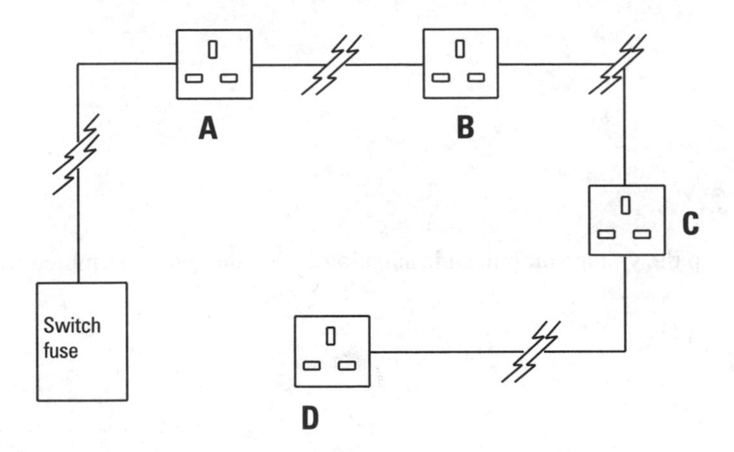

Figure 2.17

Suggested Time: 1 hour

Equipment: The equipment consists of 4 × 13 A socket outlets wired in 1 mm^2 cable 3 m apart. A pair of shielded connectors can be used at each outlet for voltage readings.

5 A load
10 A load
100 W lamp

Method:

1. Connect up the circuit shown to a 230 V a.c. supply and measure the voltage and current at each socket outlet position.
2. Plug a 5 A load into socket outlet **A** and measure the voltage and current at each socket outlet position.
3. Plug a 10 A load into socket outlet **D** leaving the 5 A load in each socket outlet **A**. Measure the voltage and current at each socket outlet position.
4. With the previous two loads connected plug a 100 W lamp into socket outlet **C**. Measure the voltage and current at each socket outlet position.

Results:

	A		**B**		**C**		**D**	
	V	*I*	*V*	*I*	*V*	*I*	*V*	*I*
1								
2								
3								
4								

WARNING!

This exercise should be carried out in accordance with the Electricity at Work Regulations 1989 and all live conductors should be shielded and insulated from touch. All test probes should meet the requirements of the Health and Safety Executive Guidance Note GS38.

Questions:

1. What was the maximum percentage voltage drop in the tests in Task 10?

2. Why is it important to keep the voltage drop in an installation within the figure permitted by Reg. 525-01-02?

Measurement of power

In a simple d.c. circuit, the power can be measured by multiplying the readings obtained from an ammeter and a voltmeter.

$$P = U \times I$$

In an a.c. circuit, this is no longer reliable since the current and voltage may no longer be in phase with each other and the result obtained will be the VoltAmps rather than the true power of the circuit.

The electrodynamic (dynamometer) wattmeter connected as shown in Figure 2.18 will read the product of current and voltage at the same instant and will indicate the true power of the circuit.

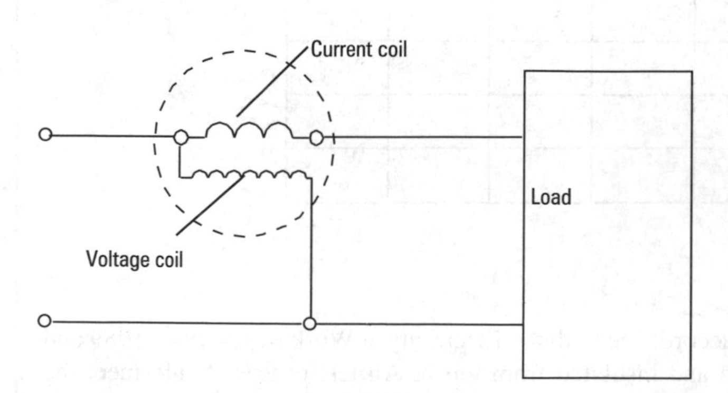

Figure 2.18 *Circuit diagram of wattmeter*

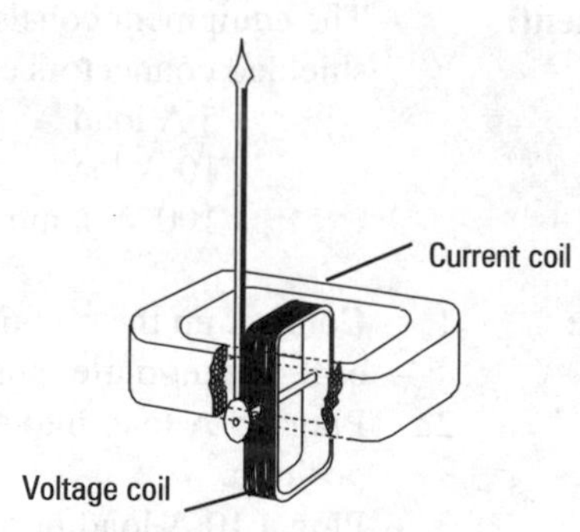

Figure 2.19 *Dynamometer wattmeter*

As a portable instrument, the wattmeter has to be suitable for a wide variety of currents and voltages. If, for example, a wattmeter has three voltage and four current ranges this means that the deflection will have twelve possible interpretations.

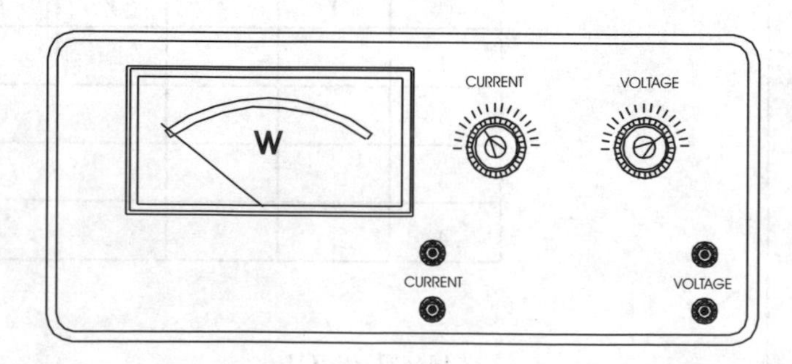

Figure 2.20 *A practical meter with voltage and current connections*

A scale of such complexity is obviously impractical so a single scale will be used and this will be read and multiplied by a factor determined by the current and voltage ranges selected.

Example

A wattmeter reads 0.7 on a scale of 0 to 1 when the current range is 0–1 A and the voltage range is 0–100 V If the multiplying factor accompanying this choice of ranges is 100 then the power indicated is

$$0.7 \times 100 = 70 \text{ Watts.}$$

If however the instrument indicated 0.35 on the 0–1 A and 0–200 V ranges where the multiplying factor is 200 then the power indicated would be
0.35 x 200 = 70 Watts.

Wattmeter connections

As power is the product of voltage and current it follows that a wattmeter is basically a voltmeter and ammeter combined. Providing the voltage coil is connected as a voltmeter across the potential, and the current coil is connected as an ammeter in series, there should be no problems using these. However, should the windings be incorrectly connected serious damage can be caused to the meter and the circuit.

There are usually 4 terminals on a wattmeter, 2 for the voltage coil and 2 for the current.

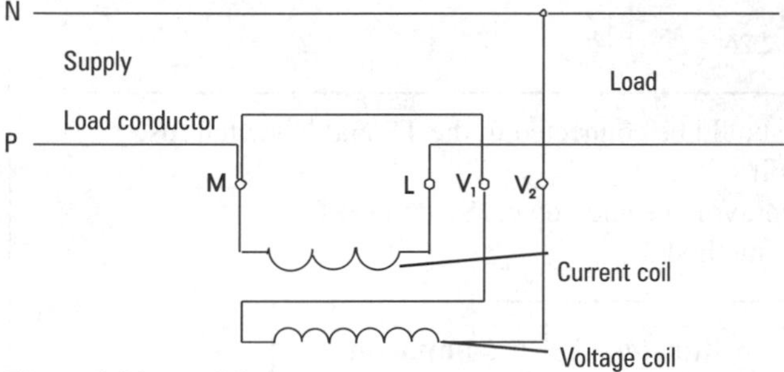

Figure 2.21 *Meter connections*

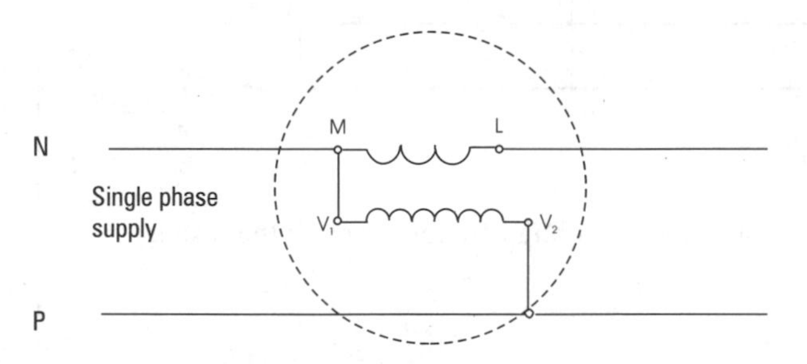

Figure 2.22 *Circuit diagram*

Measuring power in three phase circuits

Where the power is to be measured in three phase circuits this can become more complex. If, however, the loading on each phase is the same this can be carried out using just one wattmeter.

The wattmeter is connected between one phase and the star point as though it is a single phase supply. The reading on the wattmeter can now be multiplied by three to find the total power of the load. This method can only be used on balanced three phase loads where the star point is available.

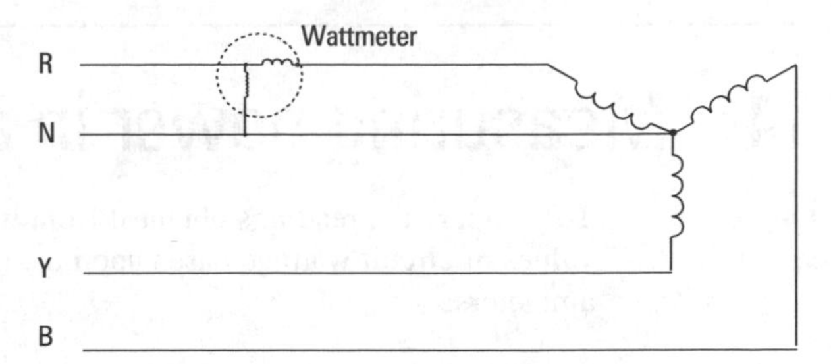

Figure 2.23

The reading obtained by a wattmeter is the true power in the circuit and if this is at all reactive the power will not be equal to the voltage times the current. By taking voltage, current and power readings of the same load its power factor can be calculated from

$$\text{power factor} = \frac{\text{true power (wattmeter reading)}}{\text{voltage} \times \text{current}}$$

The answer should always be less than one.

Measuring voltage, current and power on high voltage/current a.c. systems

Figure 2.24 shows the use of instrument transformers for metering purposes on high voltage and high current a.c. systems.

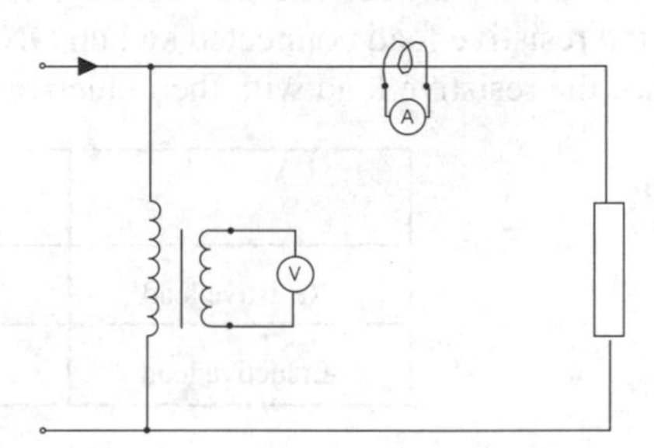

Figure 2.24

Figure 2.25 shows the connection of a single-phase wattmeter used to measure the power of a high voltage and high current a.c. system.

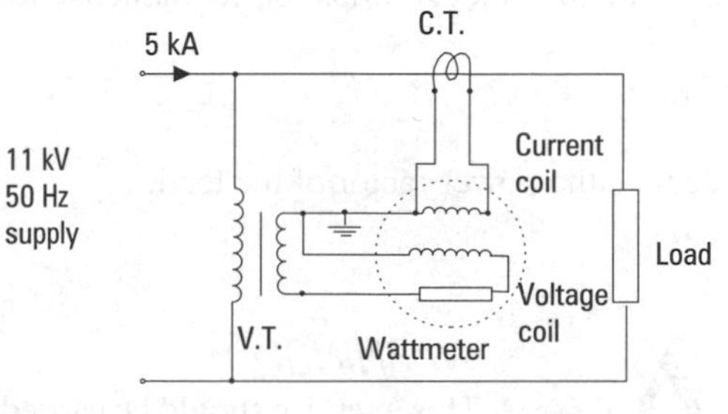

Figure 2.25

The current and voltage transformer "isolate" the wattmeter from the high current/voltage system.

11 Measuring power in a.c. circuits

Objective: To compare the readings obtained from wattmeters with the calculated values of circuit wattage based upon the readings of voltmeters and ammeters.

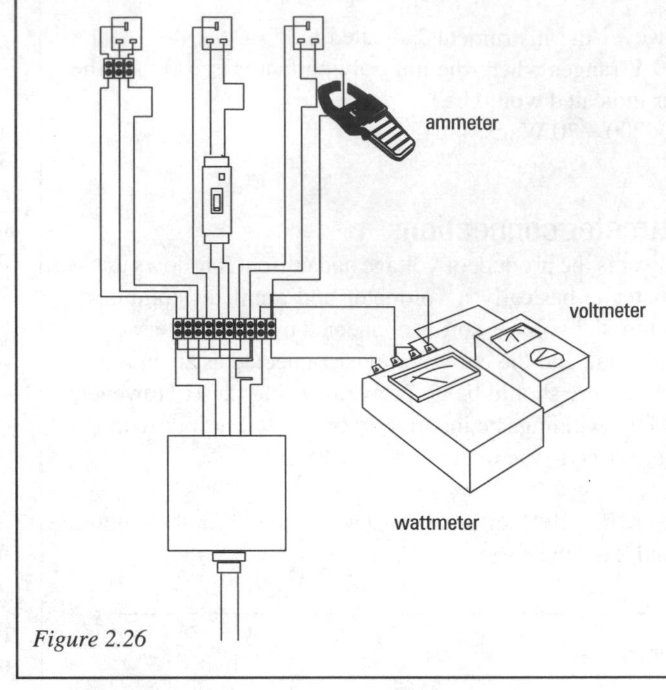

Figure 2.26

Equipment:
1 simulated distribution system
1 wattmeter 0–1000 W
1 ammeter (clamp on)
1 voltmeter 0–230 V
1 resistive load approximately 1 kW
1 inductive load (for example a fluorescent fitting without power factor capacitor)

Suggested time: 1 hour

Method:

1. A three phase and neutral supply with earth leakage protection should be connected to the TP and N switch fuse.
2. Connect up the wattmeter removing the LINK for the current coil.
3. With the resistive load connected switch ON and note the current, voltage and power. Switch **OFF**.
4. Replace the resistive load with the inductive load and repeat the method 3.

Results:

	wattmeter	voltmeter	ammeter
Resistive load			
Inductive load			

Questions:

1. Calculate the volts × amps for the resistive load, compare it to the wattmeter reading and comment on the result.

2.a) Calculate the volts × amps for the inductive load, compare it to the wattmeter reading and comment on the results.

b) Calculate the power factor of the load.

WARNING!
This exercise should be carried out in accordance with the Electricity at Work Regulations 1989 and all live conductors should be shielded and insulated from touch. All test probes should meet the requirements of the Health and Safety Executive Guidance Note GS38.

3

Selecting Appropriate Wiring Systems

Before starting this chapter it would be helpful to read through again the Chapters 4 and 5 in the intermediate book "Stage 1 Design".

On completion of this chapter you should be able to:

- select an appropriate wiring system for a set of information including the use of building, environmental conditions, possible load currents and methods of overcurrent protection available

Introduction

The wiring system is the heart of any installation. If the wiring breaks down the loads fail to work, and if the insulation fails dangerous conditions can arise. Selecting the correct system for the load, environment and cost is very important. Having carried out a basic selection it is then necessary to confirm the suitability by calculation.

Cable selection

The selection of a particular type of cable for an installation depends on many factors. These can include environmental conditions such as heat, damp, corrosive atmospheres and mechanical damage; or specification requirements that insist on low quantities of smoke, toxic and corrosive gases should a cable catch fire.

The actual selection is often a compromise between what is suitable and what is financially acceptable. For example mineral insulated metal sheathed cable (MIMS) is very suitable for wiring domestic premises but the cost would be many times greater than if PVC insulated and sheathed cable is used. There are occasions where cost must be given a low priority and safety given the maximum consideration. When supplying the pumps of a petrol filling station, for example, it would be extremely dangerous to use PVC insulated and sheathed cable. In fact it would also be against all safety recommendations which only allow for PVC sheathed MIMS and PVC/SWA/PVC cables. PVC in these cases can be substituted with other compounds such as XLPE (cross-linked polyethylene).

Regulations also dictate the type of wiring system in some special installations. In Part 6 of the IEE Wiring Regulations under Agricultural and Horticultural Premises where it is expected that livestock may be present, the electrical equipment should be of Class II construction, or constructed of or protected by suitable insulating materials. Despite the fact that the environment is damp and open to mechanical damage, this rules out the use of wiring systems such as single cables installed in galvanised steel conduit.

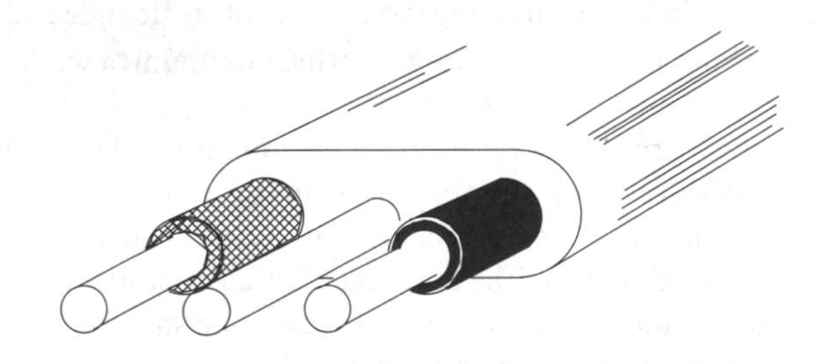

Figure 3.1 *PVC twin and earth cable*

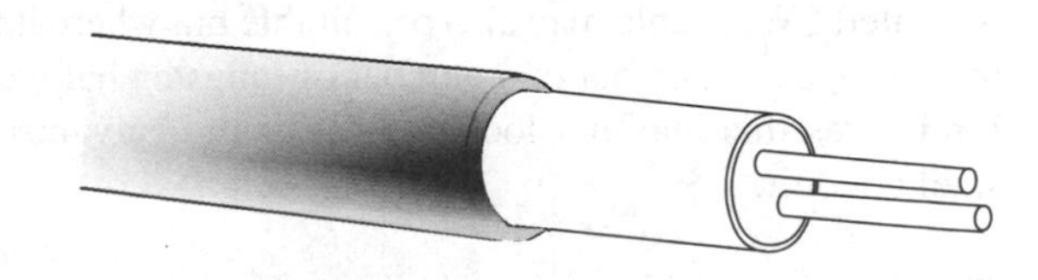

Figure 3.2 *MIMS cable*

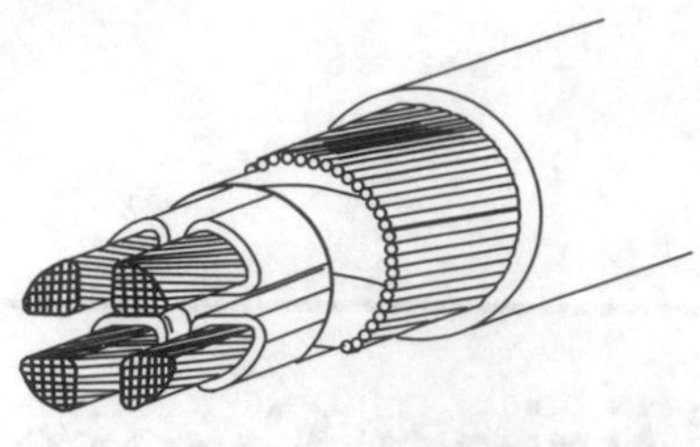

Figure 3.3 *SWA cable*

Example

Give examples of the wiring systems that would be suitable for the following installations.

(a) A domestic installation for lighting and socket outlets.
(b) An outside lighting installation which comprises of a number of tungsten halogen luminaires mounted on an outside wall. All cable runs must be on the outside surface.
(c) A fire alarm system.
(d) A motor driving a compressor.

Give reasons for your answers.

Answer

(a) Domestic installations would normally be wired in PVC insulated and sheathed cable. This would need to be protected from mechanical damage in some situations and run in positions where it is not exposed to ambient temperatures exceeding 30 °C.

The reasons for this choice are that:

- it is comparatively cheap compared with other systems
- no special tools are required to terminate it
- the installation can be quite fast as the cable is flexible and
- it can be concealed from view with the minimum of work.

(b) There are several wiring systems that should be considered for the installation of outside lights. Whichever system is chosen it must be capable of withstanding all of the different weather conditions. This may exclude the use of PVC insulated and sheathed cable depending on the exact location.
If the wiring system is to be exposed to any mechanical damage then a galvanised conduit system with single cables installed in it may be the best solution. PVC insulated SWA cable may also be suitable but where it is to be looped in and out of a number of tungsten halogen luminaires this may not look as neat as the galvanised steel conduit.

(c) The choice of wiring system for a fire alarm system may depend on the installation requirements but in general, MIMS or FP200 cable would be used. It is important that fire alarm systems carry on operating as long as possible in the event of a fire. Both MIMS and FP200 cables are capable of operating even when they are exposed to direct flames.

(d) Where a motor is driving a compressor there is going to be vibration. This means the wiring system chosen has got to be capable of withstanding this vibration without deterioration. Although MIMS cable can be used for this, flexible metal conduit with single-core PVC insulated cables is often more practical. It must be remembered that where flexible conduit is used, whether metal or not, a separate circuit protective conductor must be installed.

Try this

What factors should be considered when selecting a wiring system for a paint store?

Tip:
Remember that a paint store may contain fumes that could become a hazard.

Load calculations and conductor selection

It is important to ensure that any cable installed is capable of safely carrying the maximum load current in all conditions. Sometimes a total load is made up of several small loads. In this situation it must be determined what chances there are of all the loads being on all at the same time. Take, for example, the lighting in a domestic installation. It is very seldom that every light is on together. This is known as giving consideration to diversity and Table 1B in the IEE On Site Guide shows that you need only allow for 66% of the total current demand for the lighting load in a domestic installation. If, of course, you know this figure will be exceeded then the higher value must be allowed for.

Often the loading of equipment is given as the power in watts and this has to be converted into current in amperes before any cable capacity tables can be used.

Remember:

$$P = UI$$

so

$$I = \frac{P}{U}$$

There are other factors that must be taken into account before the tables can be used. It pays to follow a set sequence when doing this so that nothing is missed out:

1. Calculate the load current – where necessary take any power factor and efficiency into account.
2. Select an appropriate protection device – it must not be rated less than the load current.
3. Take any other conditions into account
   - ambient temperature
   - grouping
   - thermal insulation
   - type of protection device

Example

A single-phase motor is rated at 3 kW at a power factor of 0.85 when supplied with 230 V. The motor is protected by a fuse to BS 88. The PVC insulated cables are installed in conduit with one other circuit and the highest ambient temperature is 35 °C.

1. Calculate the load current

$$P = UI \cos \phi$$

As we need "*I*"

$$I = \frac{P}{U \cos \phi}$$

$$= \frac{3000}{230 \times 0.85}$$

$$= 15.34 \text{ A}$$

2. Select an appropriate protection device. The nearest fuse without being less than 15.34 A for a BS 88 is 16 A.
3. Take any other considerations into account.

Ambient temperature

To find the correct factors for this Table 4C1 is used. As we are using general purpose PVC cable the correction factor

for 35 °C is **0.94.**

Grouping

Before Table 4B1 can be used the reference method for a conduit installation must be established from Table 4A. We assume that the conduit is on the surface so Method 3 applies. The number of circuits we have is ours plus one other so from Table 4B1,

Reference Method 3, 2 circuits
– a correction factor of **0.8.**

Thermal insulation

As our installation does not encounter thermal insulation we use the factor of 1. If we needed to take this into account Table 52A and Regulation 523-04-01 would apply.

No thermal insulation – Factor **1**

Protection device

This only applies to the use of BS 3036 devices where a factor of 0.725 must be used. As we are using a device to BS 88 we can use the factor of **1**.

Now we can calculate the current to use when looking at the cable current rating tables.

For this calculation we must use the current that is the rating of the protection device as it is this current the cable will carry until the device operates.

$$\frac{\text{current rating of device}}{\text{factor for ambient temperature} \times \text{factor for grouping} \times \text{factor for thermal insulation} \times \text{factor for protection device}}$$

or

$$\frac{I_n}{C_a \times C_g \times C_i \times C_f}$$

$$\frac{16}{0.94 \times 0.8 \times 1 \times 1}$$

21.28 A

The next stage is to find a cable capable of carrying the current of 21.28 A. This of course means the cable rating in the tables must not be less than 21.28 A.

There are again a number of stages to go through.

i) Select the correct table
   – core arrangement and
   – type of insulation
ii) Select the appropriate columns
   – Reference method
   – number of cores
   – single or three-phase

Using the Tables in Appendix 4 of the Wiring Regulations or Appendix 7 of the On Site Guide the appropriate Table for our example can now be selected. This is Table 4D1A or Table 7A1 for single core PVC cables. The columns that apply are those under Reference Method 3 and in particular Column 4 which applies to 2 cables single-phase. In this column the first cable rating above 21.28 A is 24 A and from column 1 we can see that this is for a 2.5 mm^2 conductor.

There is one more stage that must be checked before we can be sure this conductor size is suitable. This is to make sure that when the motor is working on full load the voltage drop is not such that the motor would not be able to work correctly. Before we can determine the voltage drop there are two more details we need to know about our motor circuit. These are, the length of cable run and the minimum voltage that the motor will work on. The length of run we will assume is 10 metres and the minimum voltage is 225 V, giving, a maximum voltage drop of 5 V on full load.

Details of the voltage drop for a 2.5 mm^2 cable for Reference Method 3 can be found on Table 4D1B or 7A2 col 3. This is 18 mV for a 2.5 mm^2 cable for every ampere that flows through every metre of cable. We have 15.34 A and 10 m, so the actual voltage drop is

$$\frac{18 \times 15.34 \times 10}{1000} = 2.76 \text{ V}$$

As this is under the 5 V maximum we find the 2.5 mm^2 conductor under these conditions is acceptable.

Although this does appear to be a lot of work in looking different things up it is important to complete all stages in the correct order. It can be dangerous not to check all the factors before a cable is installed.

Remember

If you have been given a free hand in the choice of cable, the following factors need to be considered:

Environmental
The cable must be suitable for the environment.
Is the cable of your choice suitable for:
  the ambient temperature
  the presence of moisture
  the presence of corrosive or polluting substances
  mechanical impact or vibration

Current-carrying capacity
Is the cable of your choice suitable for:
  the anticipated design current, given the type of overcurrent protection
  the voltage drop requirement
  performance under fault conditions?

Economics
Is the cable of your choice the most economic given the cost of the cable, the cost of installation and the prospective life span of the installation?

Example

Three electric heaters are to be installed in a workshop. Each heater is rated at 3 kW, 230 V and is to be wired with a separate PVC steel wire armoured cable. The three cables are run together, clipped direct to the surface, for a large part of their 15m length and go through an area with an ambient temperature of 30 °C. To ensure the maximum voltage drop is not exceeded no circuit should be greater than 6 V.

(a) Determine the design current for each circuit and the rating of each protection device if BS 88 fuses are to be used.
(b) Calculate the minimum current rating of each cable.
(c) Select a minimum size of cable to comply with the current carrying capacity and voltage drop constraints.
(d) Determine the actual voltage drop when the cable is carrying the full 3 kW load.

Answer

(a) Design current of each circuit

$$= \frac{3000}{230}$$

$$= 13 \text{ A}$$

The protection device to BS 88 for 13 A = 16 A

(b) The minimum current rating of each cable

$$\frac{\text{Protection device rating}}{\text{correction factors}}$$

Relevant correction factors:

grouping – 3 circuits touching = 0.79

temperature = 30 °C = 1

$$\frac{I_N}{C_a \times C_g}$$

$$= \frac{16}{1 \times 0.79}$$

$$= 20.25 \text{ A}$$

(c) Minimum cross sectional area of conductor

Table 4D4A col 2 Table 4D4B col 3

1.5 mm^2 21 A 29 mV

A 1.5 mm^2 conductor is large enough for the current rating.

Checking for voltage drop

length of run = 15 m

current = 13 A

$$\text{voltage drop} = \frac{29 \times 15 \times 13}{1000} = 5.66 \text{ V}$$

(d) The actual voltage drop is 5.66 V.

Try this

A 400 V three-phase 9 kW water heater is to be wired with PVC insulated single conductors installed in steel conduit. The water heater is 15 m from the distribution board at the main intake position. Overcurrent protection is be HBC fuses to BS 88. For 5 m of the run a second similar circuit is installed in the same conduit. Calculate the minimum cross sectional area of cable suitable if the maximum voltage drop must not exceed 5V.

Tip

Remember to look up first:

– the reference method for the installation of cables Table 4A

– grouping factor Table 4B1

Calculate the line current – remember $P = \sqrt{3}$ UI

Look up a suitable BS 88 protection device.Calculate the maximum load current for the cable:

$$\frac{\text{rating of protection device}}{\text{correction factors}}$$

Look up a suitable cable Table 4D1A

Check the voltage drop Table 4D1B

Check against 5 V.

The earth fault path

The earth fault path consists of the circuit protective conductor and the earth resistance external to the installation.

To complete the circuit the phase conductor is also included in the resistance path.

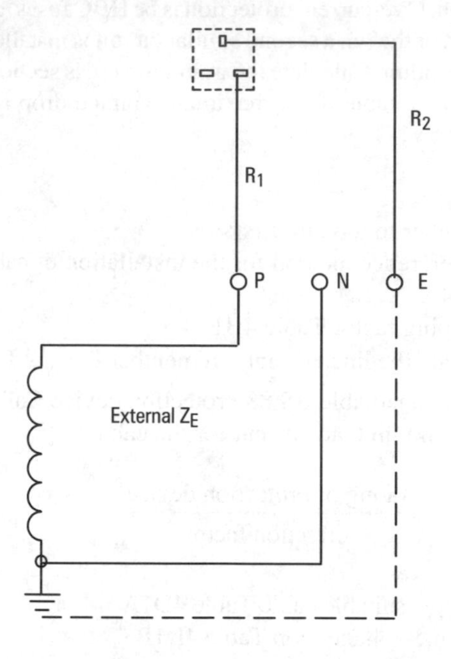

Figure 3.4 $Z_s = Z_E + R_1 + R_2$ *where* Z_s *= complete earth fault path*

As the circuit external to the installation is in the hands of the supply company this is all put together as one value covering the impedance of the earth circuit, transformer winding and phase conductor. This impedance is represented by Z_E, the external impedance.

The cross sectional area of the phase conductor is determined, as we have seen, by the current demand of the load. The resistance of this is referred to as R_1.

The cross sectional area of the circuit protective conductor cannot be determined in the same way as the live conductors, for it only comes into use under fault conditions. Then it has to help clear the fault current as fast as possible.

The maximum time that a fault can be cleared in depends on the type of circuit and where it is installed. In some circumstances it is safe to have faults flowing for up to 5 seconds. In other situations it is essential to have them cleared in 0.2 seconds. As the resistance of the phase conductor has already been determined and the external impedance is outside of our control, it is often the resistance of the circuit protective conductor that is used to determine the value of prospective earth fault current.

To keep the selection of cpcs simple Table 54G in the Regulations gives a quick guide. Assuming the circuit protective conductor is copper, the same as the phase conductor, then for circuits with a phase conductor no greater than 16 mm^2 the cpc would be the same size. When the phase conductor is between 16 mm^2 and 35 mm^2 the cpc need be no greater than 16 mm^2. For circuits where the phase conductor is greater the 35 mm^2 the cpc need be no larger than half that of the phase. In practice this will need rounding up to the next actual size available. Like most quick and easy guides the answer obtained is not always the most cost effective. The same applies with this. To get a more accurate calculation Regulation 543-01-03 should be consulted and the formula given there, applied.

$$S = \frac{\sqrt{(I^2 t)}}{k}$$

where

- $S =$ the cross sectional area of the conductor
- $I =$ the current required to flow under fault conditions
- $t =$ the time the protection device will take to operate when the fault current is flowing
- $k =$ the relevant factor found in Tables 54B to 54F.

This formula takes into account the fact that heat is produced when a fault current flows and this heat can damage the insulation of the cable if it is not cleared within the appropriate time.

Example

Determine the minimum cross sectional area of protective conductor from

$$S = \frac{\sqrt{(I^2 t)}}{k}$$

for a 230V single-phase circuit which has the following:

i) a value for Z_E of 0.26 Ω.
ii) a value for $R_1 + R_2$ of 0.7 Ω
iii) a circuit protective device of 30 A to BS 1361.
iv) a k factor of 143.

Answer

To calculate the minimum cross sectional area of protective conductor the following formula must be used.

$$S = \frac{\sqrt{(I^2 t)}}{k}$$

"I" can be calculated from

$$I_a = \frac{U_0}{Z_s}$$

where

- I_a = current under fault conditions
- U_0 = supply voltage
- Z_s = complete earth fault loop impedance

Zs can be calculated from

$$Z_S = Z_E + R_1 + R_2$$

$$Z_E = 0.26\ \Omega$$

$$R_1 + R_2 = 0.7\ \Omega$$

so $$Z_S = 0.26 + 0.7 = 0.96\ \Omega$$

$$I_a = \frac{U_0}{Z_S} = \frac{230}{0.96} = 240\ \text{A}$$

t can be determined from the time/current characteristics for a 30A BS1361 fuse shown in Fig.1 Appendix 3 of the Regulations.

t = 30 A fuse 240 A fault current
= 0.2 seconds

$$\therefore s = \frac{\sqrt{240^2 \times 0.2}}{143}$$

$$= 0.75\ \text{mm}^2$$

This has to be rounded up to the nearest size of cable, which is 1.0 mm².

Try this

A 230V circuit is wired in a 6 mm² PVC/SWA/PVC cable and is protected by a 50 A BS 88 general purpose fuse.

(a) What is the maximum value of Z_S this circuit can have if the disconnection time must not exceed 0.4 seconds?

(b) Determine the maximum earth fault current.

Tip:

Remember Z_S values can be found in the IEE Regulations in Tables 41B1, 41B2 and 41D; and Tables 2A, 2B and 2C in the On Site Guide Regulation 413-02-08 can help with I_a.

The final value of Z_e must be adjusted to allow for temperature rise in the conductors.

Fire detection devices

It is important to consider the devices that are used on fire detection circuits. These often require special circuits that give the maximum protection in the event of a fire. The detection devices generally fit into three categories, smoke, heat and manual. The first two are, of course, automatic and can detect fire for 24 hours a day throughout the year, whereas the third requires somebody being present and setting off the alarm. There are other automatic systems in use but these usually require specialist installation.

Smoke detectors, as the name implies, detect the presence of smoke in the air. There are generally two types in use, the ionised chamber and the optic chamber. The use of both is very similar and it is the ionised chamber type that has become widely used in domestic installations. They should not be used where there is the possibility of smoke or dust in the air under normal working conditions as this can set them off accidentally.

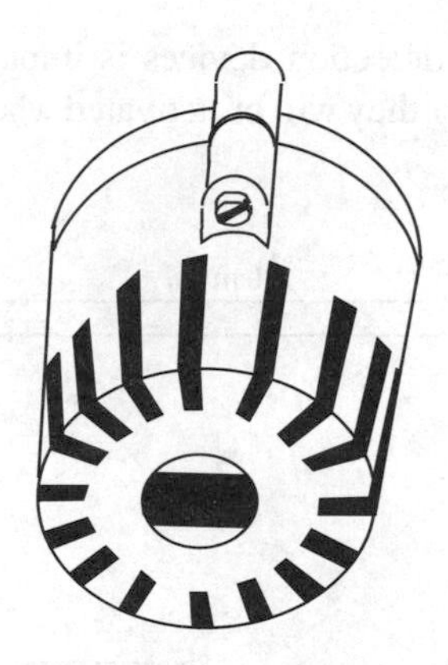

Figure 3.5 *Smoke detector*

Heat detectors also fall into two general types, those that use bimetallic arrangements and rely on a movement taking place to make or break contacts, or the ones that operate with the action of a thyristor.

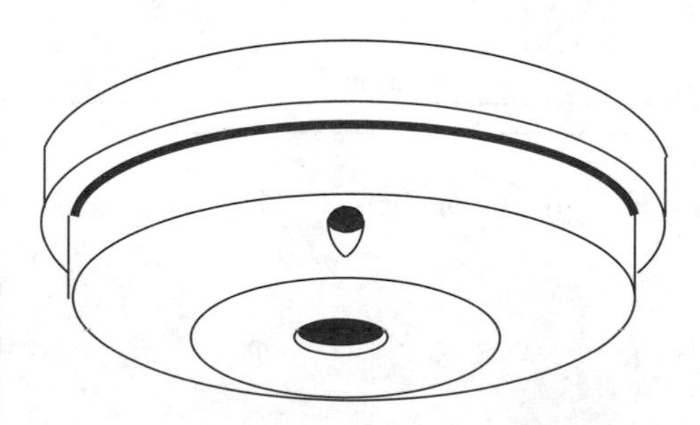

Figure 3.6 *Heat detector*

The bimetallic type have sometimes been prone to problems due to mechanical defects. Both can be affected by the sudden rise in temperature which is not necessarily anything to do with a fire. "Rate of rise" detectors can take into account changes in temperature within an area without setting off alarms.

Manual units such as break glass switches can be effective as the human body can often detect fire as soon, if not before, automatic devices. They are however prone to abuse by being set off when there is no fire and being inaccessible when there is one.

Figure 3.7 *Manual unit*

The siting of all detection devices is important as this may determine whether they will be activated when the need arises. Figures 3.8 and 3.9.

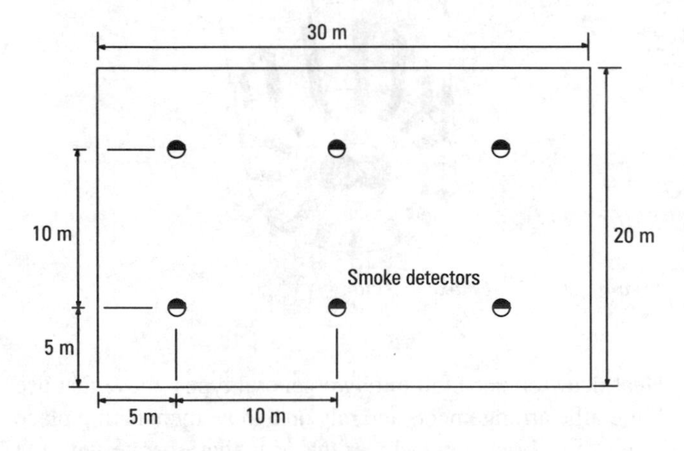

Figure 3.8 *Siting of smoke detectors on a flat ceiling*

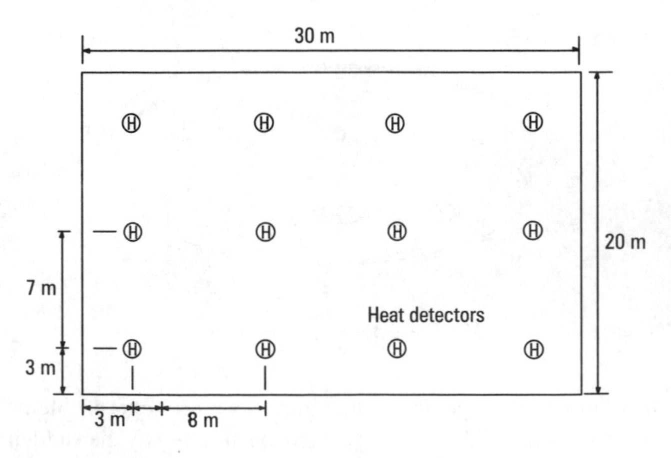

Figure 3.9 *Siting of heat detectors on a flat ceiling*

Example

A fire detection system is to be installed in an office complex. Name a suitable fire detector device for each of the following situations and give reasons for your choice.

(a) general office areas
(b) staff canteen
(c) adjacent to fire exits
(d) computer rooms

Answer

(a) The fire detection within a general office may depend on whether smoking is permitted in the area. If smoking is permitted and the area is likely to become very smoky in normal use then heat detectors would be best. If, however, the office is a smoke free zone then a smoke detector would probably be advisable. This is due to the fact that many offices use a quantity of electrical equipment and it can be this when a fault develops that causes fires. When things such as waste paper baskets catch fire then either detection device would be suitable. In addition to the automatic devices, manual break glass contacts should be sited on all exit routes.

(b) The staff canteen can be a difficult problem for it is often very smoky and the temperature varies in there throughout the day. Generally a "rate of rise" type heat detector is most suitable to limit any false alarms.

(c) All fire exits should have manual activated devices as these routes are going to be used if the building is occupied.

(d) Computer rooms are generally dust free environments. The smoke detector is usually most suitable but there are often more sophisticated systems in use as well due to the nature and value of the equipment.

Try this

Explain, with the aid of a diagram, how detection devices are wired on a "closed circuit" system.

Tip:
Remember that a relay is required in this type of circuit and that BSEN 60617 drawing symbols should be used.

12 Wiring Systems

Objective: To select a suitable wiring system for the building listed and an appropriate method of over-current protection. Give reasons for your choice.

Building:
A purpose built bulk-milk tank room (adjoining a milking parlour) with tiled brick walls and concrete roof.

Suggested Time: 15 minutes

Points to be considered:

✓ has a suitable wiring system been selected?

✓ have all environmental conditions been considered?

✓ have possible load currents been taken into consideration?

✓ is the over-current protection device chosen suitable?

Answer:

13 Wiring Systems

Objective: To select a suitable wiring system for the building listed and an appropriate method of over-current protection. Give reasons for your choice.

Building:
Light engineering workshop with adjoining office block. The workshop contains 20 machines (lathes and so on) and has an open apex roof. The offices have plastered walls and ceilings.

Suggested Time: 15 minutes

Points to be considered:

✓ has a suitable wiring system been selected?

✓ have all environmental conditions been considered?

✓ have possible load currents been taken into consideration?

✓ is the over-current protection device chosen suitable?

Answer:

14 Wiring Systems

Objective: To select a suitable wiring system for the building listed and an appropriate method of over-current protection. Give reasons for your choice.

Building:
Twenty storey block of flats, the building has concrete floors throughout and stairs and lift.

Suggested Time: 15 minutes

Points to be considered:

✓ has a suitable wiring system been selected?

✓ have all environmental conditions been considered?

✓ have possible load currents been taken into consideration?

✓ is the over-current protection device chosen suitable?

Answer:

15 Wiring Systems

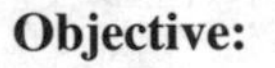

Objective: To select a suitable wiring system for the building listed and an appropriate method of over-current protection. Give reasons for your choice.

Building:
Rewiring an old house which has existing surface wiring on unplastered stone walls.

Suggested Time: 15 minutes

Points to be considered:

✓ has a suitable wiring system been selected?

✓ have all environmental conditions been considered?

✓ have possible load currents been taken into consideration?

✓ is the over-current protection device chosen suitable?

Answer:

16 Wiring Systems

Objective: To select a suitable wiring system for the building listed and an appropriate method of over-current protection. Give reasons for your choice.

Building:
A leisure centre incorporating lighting only in a sauna room, the pool area and changing rooms. There is an oil-fired boiler house.

Suggested Time: 15 minutes

Points to be considered:

✓ has a suitable wiring system been selected?

✓ have all environmental conditions been considered?

✓ have possible load currents been taken into consideration?

✓ is the over-current protection device chosen suitable?

Answer:

17 Wiring Systems

Objective: To select a suitable wiring system for the building listed and an appropriate method of over-current protection. Give reasons for your choice.

Building:
A cow shed built with steel frame-work and corrugated metal sheets.

Suggested Time: 15 minutes

Points to be considered:

✓ has a suitable wiring system been selected?

✓ have all environmental conditions been considered?

✓ have possible load currents been taken into consideration?

✓ is the over-current protection device chosen suitable?

Answer:

18 Wiring Systems

Objective: To select a suitable wiring system for the building listed and an appropriate method of over-current protection. Give reasons for your choice.

Building:
A laboratory requires
10 × 110 V BS1363 sockets,
15 × 400 V sockets and
15 × 12 V sockets for test instruments and the like.
Work benches are fitted on three walls of the room

Suggested Time: 15 minutes

Points to be considered:

✓ has a suitable wiring system been selected?

✓ have all environmental conditions been considered?

✓ have possible load currents been taken into consideration?

✓ is the over-current protection device chosen suitable?

Answer:

19 Wiring Systems

Objective: To select a suitable wiring system for the building listed and an appropriate method of over-current protection. Give reasons for your choice.

Building:
A control room for a very large food processing system which contains many three phase and single phase heaters and motors requiring separate cables to each.

Suggested Time: 15 minutes

Points to be considered:

✓ has a suitable wiring system been selected?

✓ have all environmental conditions been considered?

✓ have possible load currents been taken into consideration?

✓ is the over-current protection device chosen suitable?

Answer:

20 Wiring Systems

Objective: To select a suitable wiring system for the building listed and an appropriate method of over-current protection. Give reasons for your choice.

Building:
A petrol filling station forecourt.

Suggested Time:

Points to be considered:

✓ have suitable wiring systems been selected?

✓ have all environmental conditions been considered?

✓ have possible load currents been taken into consideration?

✓ are the over-current protection devices chosen suitable?

Answer:

21 Wiring Systems

Objective: To select a suitable wiring system for the building listed and an appropriate method of over-current protection. Give reasons for your choice.

Building:
A temporary installation with lighting and portable tool requirements for a small construction site.

Suggested Time: 15 minutes

Points to be considered:

✓ has a suitable wiring system been selected?

✓ have all environmental conditions been considered?

✓ have possible load currents been taken into consideration?

✓ is the over-current protection device chosen suitable?

Answer:

4

Motor Control and Examination

On completion of this chapter you should be able to:

- ◆ install and connect an a.c. motor with direct-on-line starter to a 230 V control system
- ◆ dismantle and reassemble a single phase capacitor start motor
- ◆ dismantle and reassemble a universal (series) motor
- ◆ dismantle and reassemble a three phase (cage or wound) motor
- ◆ connect up a number of open and closed contacts to a relay or contactor to switch a load when any contact is operated
- ◆ install and connect an a.c. motor with forward and reverse D.O.L. starter to a 400 V control system.

Introduction

Before considering the practical exercises in this chapter it may be helpful to remind ourselves of some important facts regarding motors.

There are a number of different a.c. motors that should be given consideration when dealing with this subject.

These include:

single-phase (split phase)

single-phase capacitor start

universal (series)

three-phase cage induction

three-phase wound rotor

There are several aspects to be aware of when looking at these. First it is important to have an idea of their basic construction. This does not mean every nut and bolt, but the main parts such as the type of rotor and stator. It is also necessary to know the circuit diagram of the internal connections of the windings. Again this need not be complex but simply what is connected to what, using standard symbols.

It is also important to look at the control of motors.

The workings of A.C. motors

To follow how an a.c. motor works consideration must be given to how alternating current is generated. This can be described as a magnet turning inside three coils.

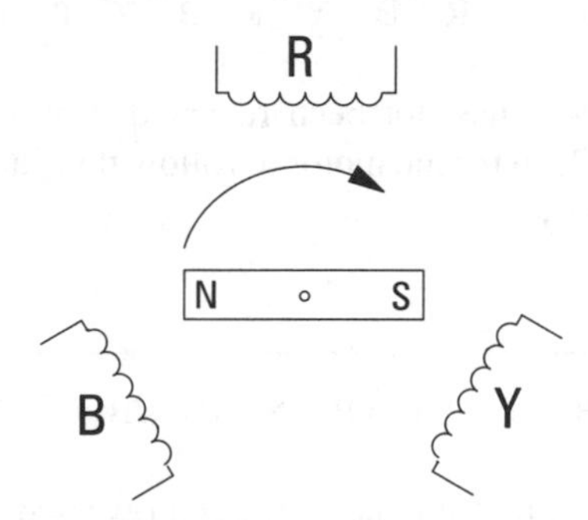

Figure 4.1

As the magnet revolves its poles go past each coil in turn. As the magnetic field is moving, an e.m.f. is induced in each coil. The coils are effectively positioned 120° apart so the e.m.f.s are induced at different times and three voltages at 120° intervals are produced.

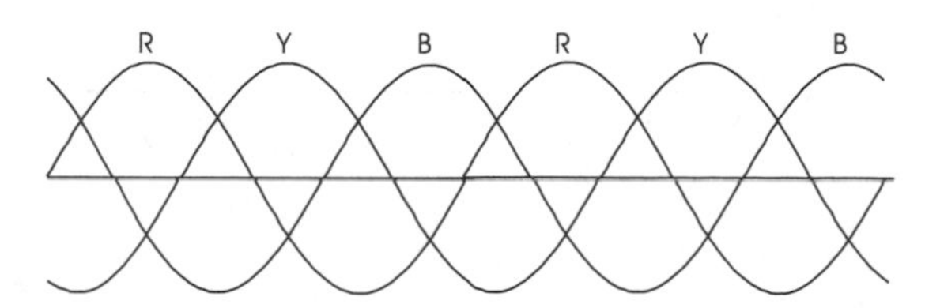

Figure 4.2

The pattern produced always follows the sequence Red, Yellow, Blue.

Now, relating this to a three-phase motor that is being supplied with the generated output, it can be considered as the inverse of a generator.

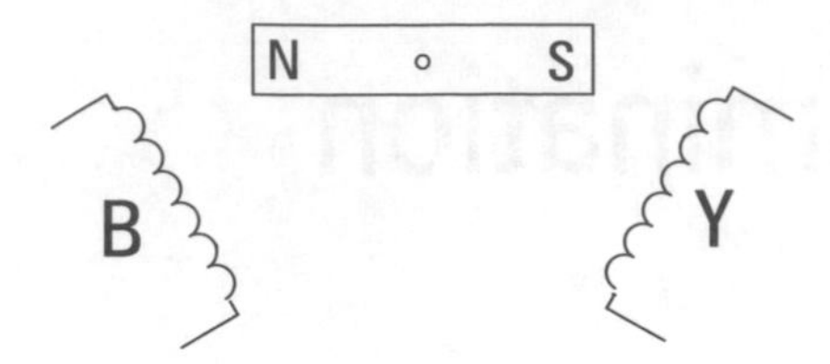

Figure 4.3

If the coils are connected up in the same sequence then the magnet will spin in the same direction. So in this case it is clockwise for the sequence RYB. If you take this out to a number of coils the sequence can become clearer.

R Y B R Y B R Y B R Y B

If any of the coils are swapped over the sequence is reversed. Changing Red and Yellow.

Y R B Y R B Y R B Y R B

As the generator has not been reversed it still follows the sequence RYB, so for the motor to follow this pattern it has to go the other way.

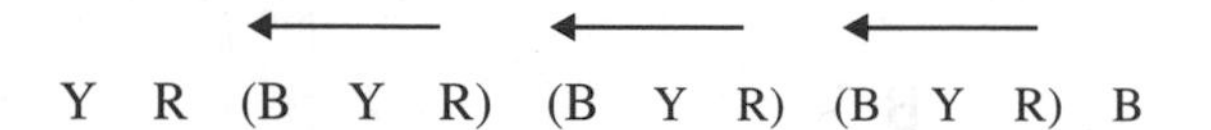

Y R (B Y R) (B Y R) (B Y R) B

The direction of rotation has now been reversed.

This works well for three-phase motors but single-phase does not have a sequence to follow. Single-phase is the output from just one of the generator coils.

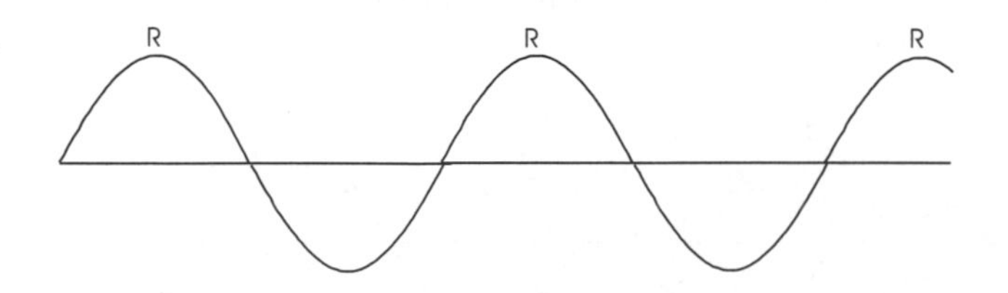

Figure 4.4

One a.c. sine wave does not create a moving magnetic field that is sufficient to start a motor. This means that a second phase has to be produced inside the machine.

There are many examples in a.c. equipment where highly inductive devices are used and a lagging phase angle is created. An inductive device consists of coils of copper wire wound around a laminated iron core. Two factors affect the total inductive effect of the device. These are the resistance of the coil and the amount of iron surrounding it. The less the resistance of the coil the greater the effect of the inductance. If the coil is deep into a laminated iron core the inductance is greater than if the coil is in air. In theory a coil with no resistance and an ideal iron core can create a phase shift of 90°. This means that the current through such a coil would occur 90° after the current through a coil with no inductance. This is, of course, also 90° behind the voltage waveform.

Now relate this to a single-phase motor. A thick conductor will have less resistance than a thin one. An example is that a 6 mm^2 cable has less resistance than a 1.0 mm^2 one. So by having two windings, one of thin wire and the other of thick, there is one with a low resistance the other with a higher resistance. As already seen a coil embedded in a laminated iron core has more inductance than one in air. So now the two coils have to be fitted into the stator. The thick coil with low resistance is to be the one that has the greater inductance so this is fitted deep into the slots on the stator. This is now almost totally surrounded by the laminated iron core. The coil with the thinner wire is fitted closer to the surface of the stator so that it has less iron surrounding it.

Now there is a coil with low resistance surrounded by a laminated iron core. When current is passed through this the inductance makes the current lag by some angle less than 90°. The second coil has a high resistance and far less laminated iron core around it so when current is passed through the coil it only lags by a small amount behind the voltage. The overall effect is that the two coils have a phase displacement between them of about 30°.

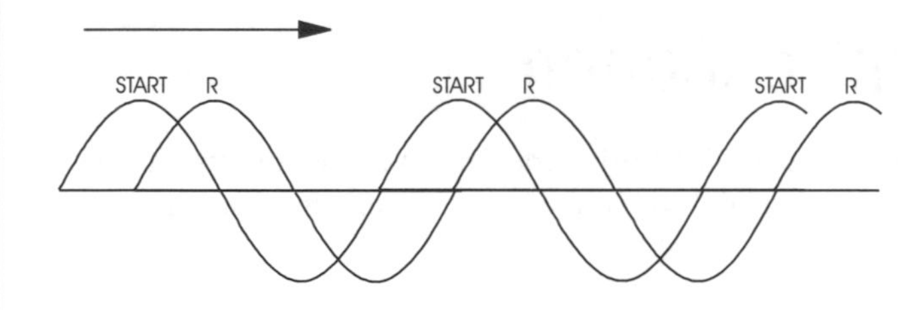

Figure 4.5

This is enough to start a motor turning. Once the motor is up to speed the two windings are no longer required and the motor will continue to run on just one winding. As the thinner of the two windings will probably not carry the load current indefinitely this is switched off. The switch for this is mounted inside the motor and automatically opens the circuit to this winding at a predetermined shaft speed. As this winding is only used to get the motor up to speed it is known as the start winding.

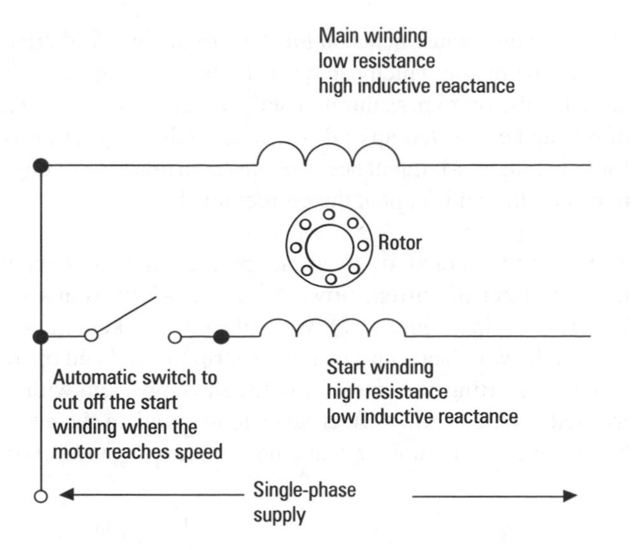

Figure 4.6

The direction of rotation of a single-phase motor depends on the direction it is started in. Assuming that the direction in Figure 4.6 is clockwise, then if the connections to the start winding are reversed the motor will rotate in an anti-clockwise direction when started.

Just changing the supply connections would have no effect on the direction of rotation because both coils have then been changed over and they are back where they started.

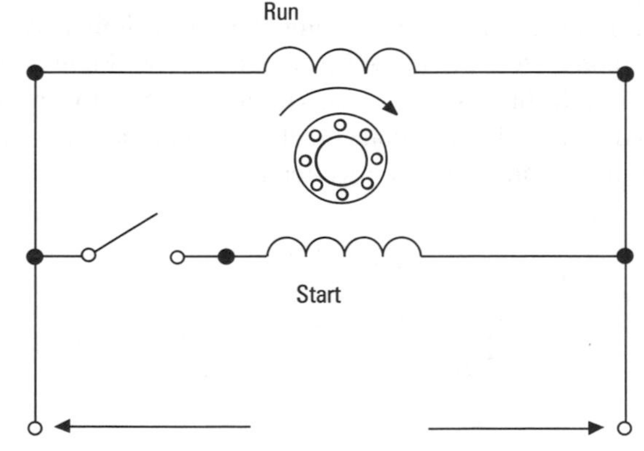

Figure 4.7

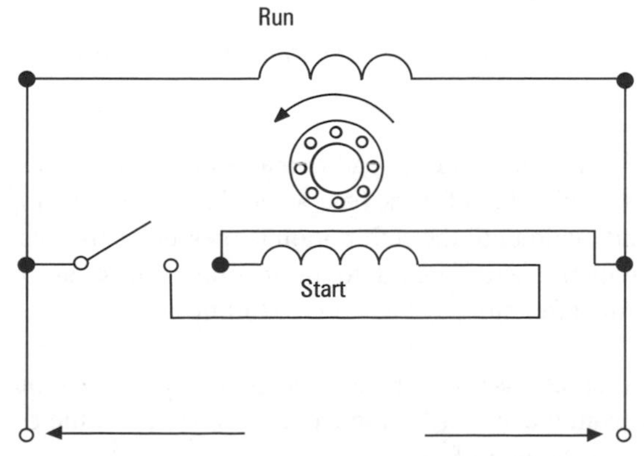

Figure 4.8 *To reverse the direction of rotation of a single-phase motor the connections to either the start or run winding can be changed over, but not both.*

To create a larger "shift" between the two windings a capacitor is used in series with the start winding. This is then known as a capacitor start split-phase motor.

The construction of a three-phase induction motor

A simplistic way to describe the action of a three-phase motor is to compare the rotor to a permanent magnet. This is in fact not true when it comes to the construction of the motor. There is no permanent magnet, in fact when the machine is switched off there is no magnetic field at all. It can get even more puzzling when you consider there is no electrical connection at all to the rotor. This means you can't put a supply on to it to make it an electromagnet. So how does it become magnetised?

The rotor of an induction motor is made up of a cage of copper or aluminium bars, as in Figure 4.9.

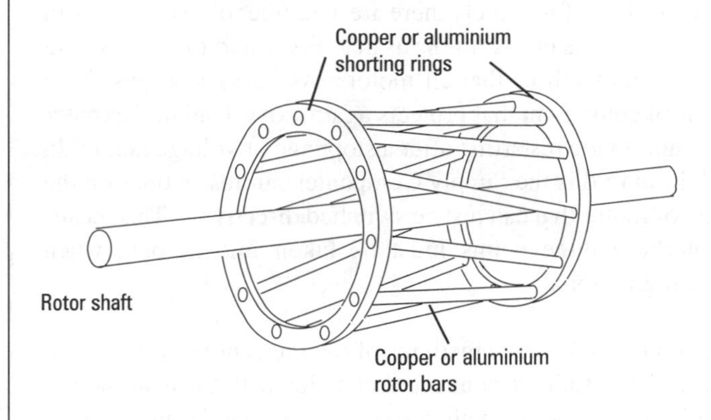

Figure 4.9 *Cage-rotor*

In practice these are not so easy to see as they are embedded in the laminated iron core. If you think of each of these copper or aluminium bars as being an electrical circuit and they are all shorted out by the end rings then we can look and see how it works.

When the motor is assembled each rotor bar is close to a coil on the stator (Figure 4.10).

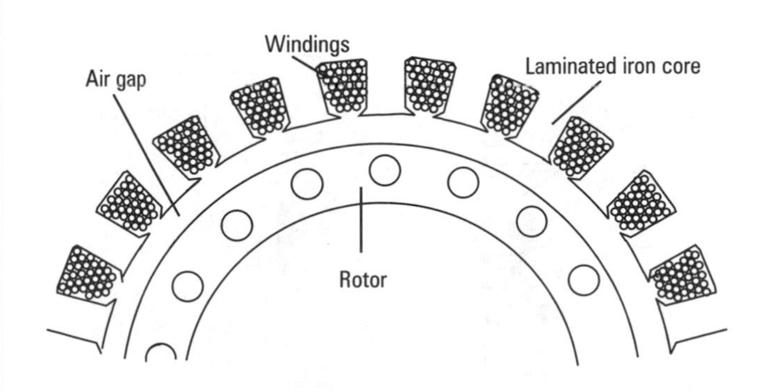

Figure 4.10 *Cross-section of an a.c. motor*

The stator coils are supplied with an alternating current which creates a moving magnetic field. This magnetic field not only goes around the coil but through the iron core across the air gap and into the rotor bars. This is the same principle as that for a double wound transformer. In this case the primary winding is the stator coil and the secondary is the rotor bars. As the rotor bars are all shorted out with the end rings the resistance of the circuit is low and the induced e.m.f. from the stator coils produces high currents in the rotor. This induced current in turn produces its own magnetic field. The magnetic field in the stator and the magnetic field in the rotor interact with each other and the rotor is moved. As the magnetic field in the stator moves with the supply voltage the rotor follows it and continues to rotate.

Motor starting

Most a.c. motors, if connected across the appropriate supply, will work. Unfortunately there are a number of reasons why in most cases it is not as simple as that. For a start the IEE Wiring Regulations tell us that all motors over 0.37 kW must have control equipment that protects against overload of the motor and automatic restarting after a stoppage or voltage failure. In addition to this the Supply Companies put restrictions on the size of motor that can just be switched directly on. This means that these three points must be taken into account when starting motors.

First let's look at two methods of overcurrent protection used in starters. Both are connected in series with the main supply cables to the motor so all current has to flow through them.

The first works on the principles of magnetism. It is basically a coil which becomes magnetised when a certain amount of current flows through it (Figure 4.11). Inside the coil is a steel plunger which is pulled through the coil when the current reaches the critical value. When it is pulled right through it operates some contacts which switch off the motor. When motors start off from a stationary start they take a lot more current than when they are running. This current would be high enough to cause the overload trip to operate. To overcome this an oil dashpot is fitted to the bottom of the plunger so that when the starting current tries to pull it the oil slows down the rate of travel. If the high current continues, as when a fault develops, the plunger will eventually pull through the oil and operate the contacts.

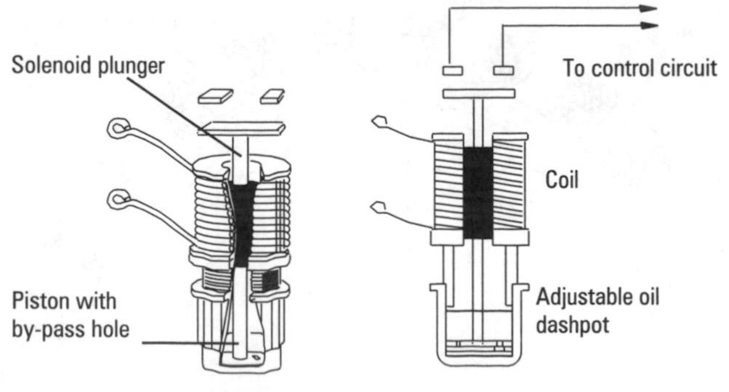

Figure 4.11 Magnetic overload

There are two adjustments on this type of device, (i) there is a hole in the piston which can be made larger or smaller, thus allowing the oil to pass through at different rates, (ii) the bath of oil can be screwed up or down altering the length of travel for the plunger. Maintenance is required on this type of device to ensure the oil is kept at the correct level.

The second method of overload protection works on the heating effect of current flow. A wire is wrapped around a bimetallic strip (Figure 4.12). When the motor takes excessive current the wire heats up causing the strip to bend and operate contacts. Starting currents are not the same problem with this method, as the heating effect takes longer and so the current has dropped to its running value before the trip can operate.

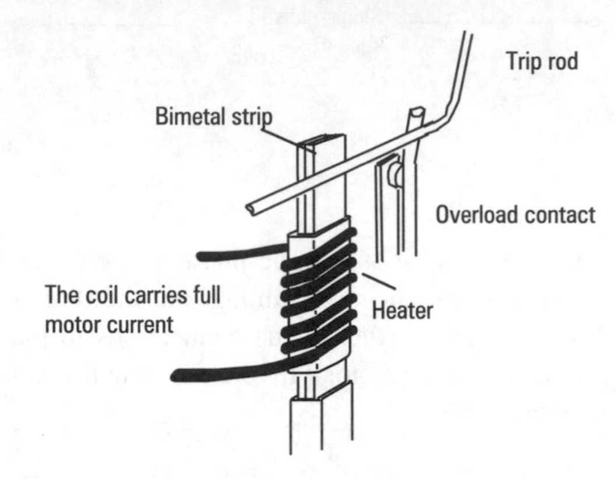

Figure 4.12 Thermal overload device

The control circuit for most motor starters is similar. A start button completes a circuit which energises a coil (Figure 4.13). The coil pulls in the main contacts to supply the motor and also an auxiliary "hold-on" contact that shorts out the start button so the motor continues to work after the start button is released.

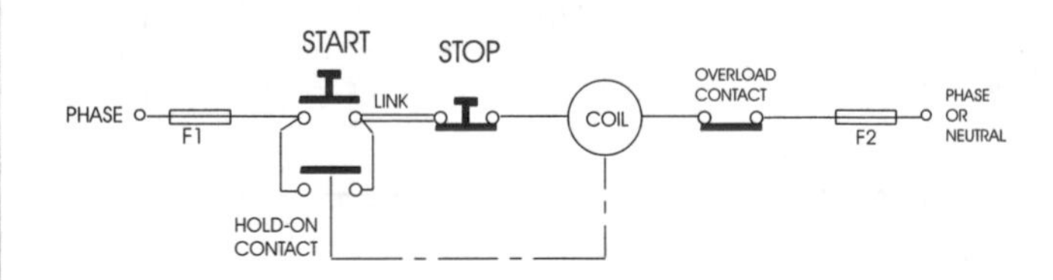

Figure 4.13 Control circuit for direct-on-line starter

The stop button and overload contacts are in series with the coil. Should either of these be opened the motor will stop and not start again until the start button is pressed. This meets the requirements with regard to protection against automatic restarting after stoppage or voltage failure.

On occasions it is necessary to have extra stop and start buttons away from the starter. These can be connected to the control circuit as shown in Figure 4.14.

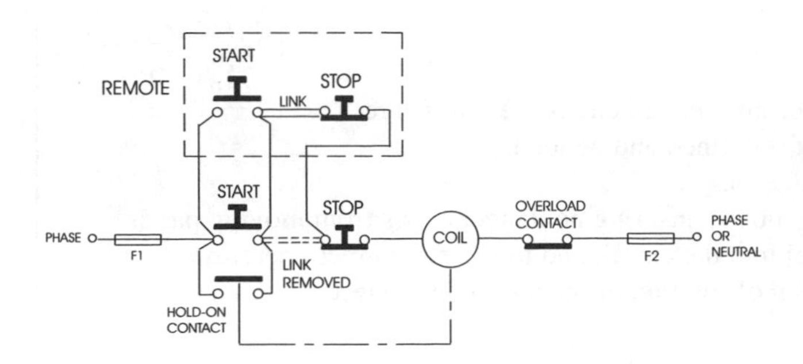

Figure 4.14 Control circuit with remote start/stop

Where a motor can be started directly onto the supply a "direct-on-line" starter is used. This is basically a three- phase contactor with the control circuits incorporated into it.

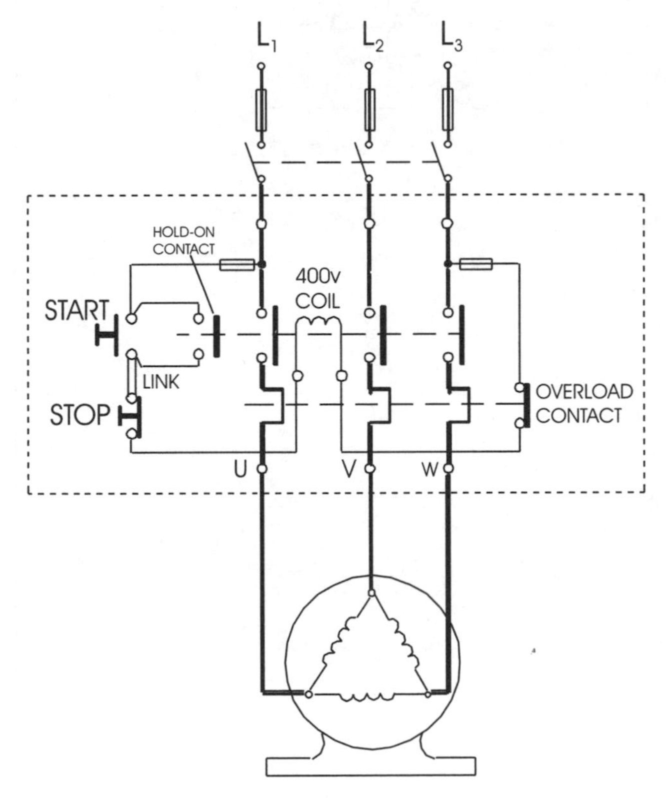

Figure 4.15 Direct-on-line starter

Power (main) circuit shown by thick lines

Control (auxiliary) circuit shown by thinner lines.

The fuses provide short-circuit protection.

The overloads may be thermally or magnetically operated.

The control circuit sometimes has to work from a 230 V supply even though it controls a 400 V three-phase motor. In this case a neutral line is required and the control circuit returns through this instead of one of the phases.

Where a motor cannot be started direct-on-line a method of current reduction must be used. The most common of these is the star-delta method of starting. For this all six ends of the motor windings must be brought out to the starter.

On starting, the stator windings are first connected in star configuration. This is to reduce the voltage across each winding. This in turn, of course, reduces the current. When the motor has got up to speed the starter is switched so that the windings are in delta configuration with the full 400 V across each winding. Figures 4.16 and 4.17 show how this can be carried out using a manual type star delta starter.

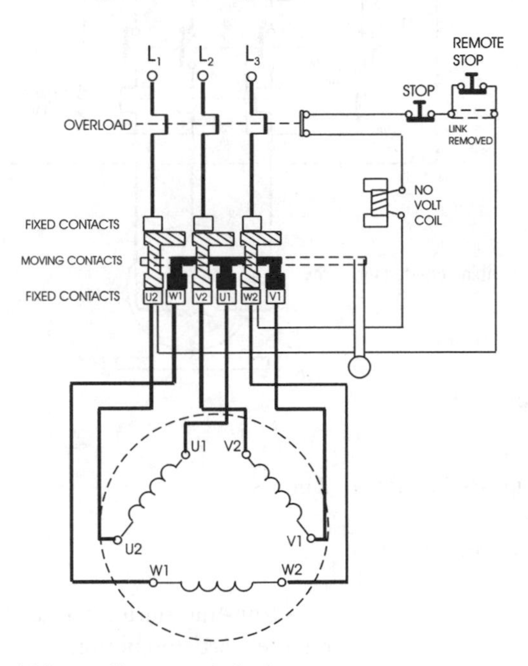

Figure 4.16 Starter switched to star

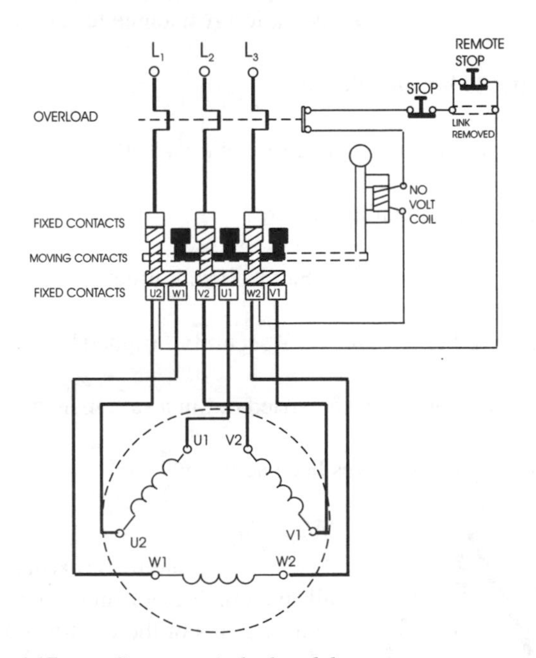

Figure 4.17 Starter switched to delta

22 Direct-on-line starter

Objectives: To wire up the starter circuit as shown, ensure that all enclosure covers are replaced and that all exposed metalwork is earthed and bonded.
To carry out appropriate inspections and tests.
To switch on the supply and test run the motor, ensuring no danger exists from moving parts.
To connect up a remote start/stop control into the circuit and to test for correct operation.
To reconnect so as to reverse the direction of rotation of the motor and to test.

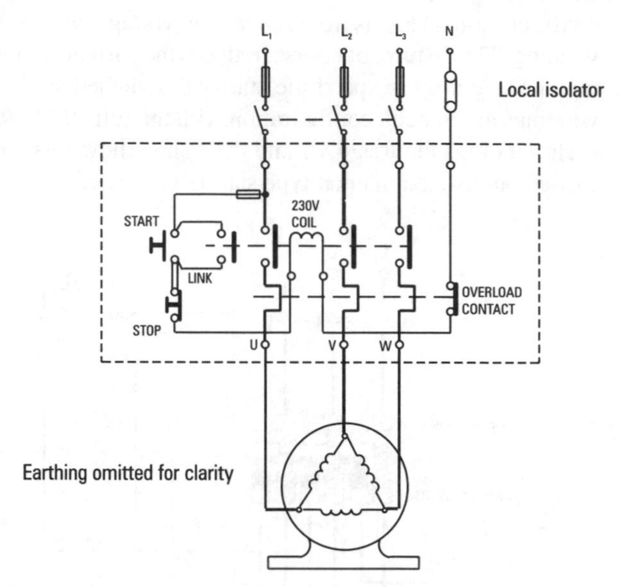

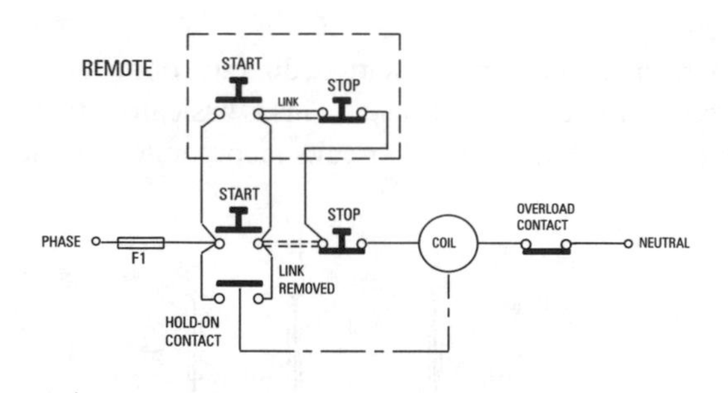

Figure 4.18

Suggested time: 2 hours

Equipment:

direct-on-line starter with 230 V coil
remote start/stop buttons
three phase cage rotor motor
insulation resistance tester

Points to be considered:

✓ has the wiring been carried out correctly?

✓ are all terminations secure?

✓ are all terminations shielded from direct contact?

✓ have all enclosures been securely fastened?

✓ has the wiring been carried out in a safe manner?

✓ have the objectives been achieved?

WARNING!
This exercise should be carried out in accordance with the Electricity at Work Regulations 1989 and all live conductors should be shielded and insulated from touch. All test probes should meet the requirements of the Health and Safety Executive Guidance Note GS38.

23 Connecting external control circuits to a single-phase contactor

Objective: To connect up a number of normally open and normally closed contacts to a contactor to switch a load when any contact is operated.

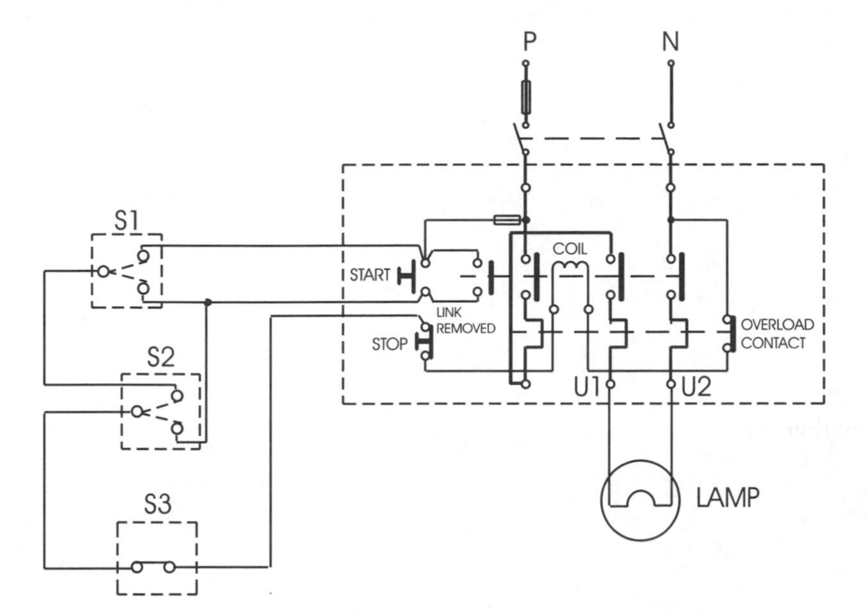

Equipment:

1 single-phase direct-on-line starter
1 60 W lamp
2 two-way switches
1 one-way switch

Figure 4.19

Suggested time: 1 hour

Equipment:

1 single-phase direct-on-line starter
1 60 W lamp
2 two-way switches
1 one-way switch

Method:

1. Test out the switches so that the contact positions are known.

2. Connect up the above circuit to a single-phase a.c. supply with all switch contacts in the positions shown.

3. From the above circuit determine how the following conditions can be achieved by using S1, S2 and S3.

Condition one

When the supply is switched on the lamp can be switched ON with the start button, and OFF with the STOP button or S3.

Condition two

The lamp comes on directly the supply is switched on and can be held OFF with the STOP button or switched OFF with S3.

WARNING!

This exercise should be carried out in accordance with the Electricity at Work Regulations 1989 and all live conductors should be shielded and insulated from touch. All test probes should meet the requirements of the Health and Safety Executive Guidance Note GS38.

Questions:

1. Draw the control circuit to achieve condition one.

2. Draw a control circuit to achieve condition two.

3. Draw the control circuit which would be used with a direct-on-line starter when 3 interlocks are used on safety guards. When any of the safety guards are removed the starter must go OFF and not be re-started until the guards are replaced.

Examination of a three-phase cage-rotor induction motor

Stage 1

Before dismantling a motor test the resistance of the three stator windings for "balance", and the insulation resistance of each winding.

Testing winding resistances for balance

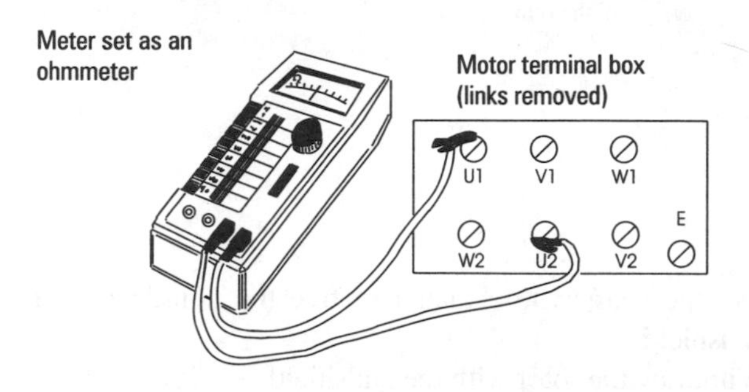

Figure 4.20 U_1 *to* U_2, V_1 *to* V_2 *and* W_1 *to* W_2 *resistances should be the same.*

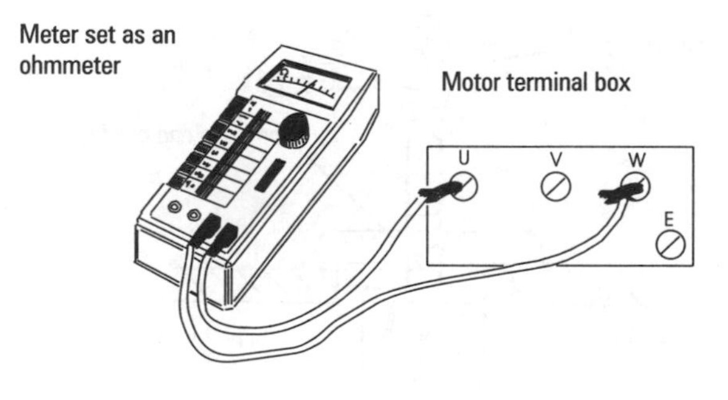

Figure 4.21 *U to V, V to W and W to U resistances should be the same.*

If one reading is much lower than the other two, then one winding could be "short circuited".

If the meter pointer deflects right across the scale this may indicate an open circuited winding or a loose winding termination.

Testing insulation resistance

This test measures the resistance between different windings and between each winding and the motor earth connection. The instrument uses a test voltage of at least 500 V so care must always be taken when carrying out this test.

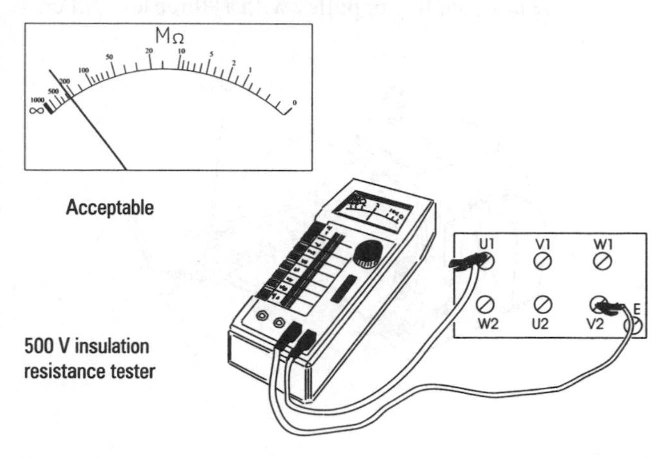

Figure 4.22

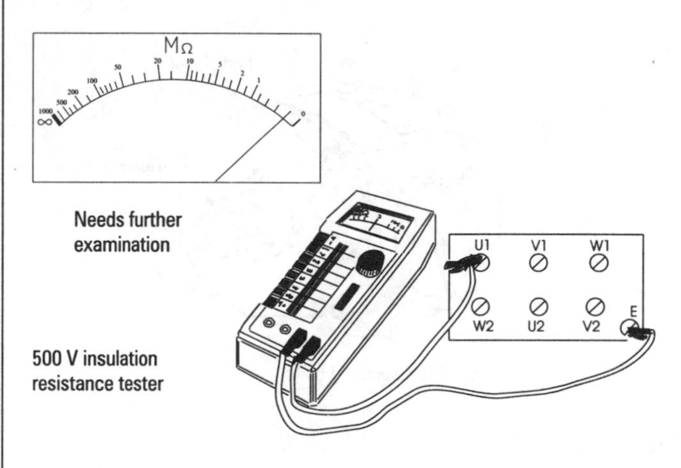

Figure 4.23

A good insulation resistance value is indicated by a high resistance reading of between 100 MΩ and infinity ∞, and a poor value is indicated by a low resistance reading (a reading of 0.5 MΩ and below is an indication of very poor insulation resistance).

Remember

Use an ohmmeter to test the resistance of the windings, and an insulation resistance tester, on the Megohms scale, to test the insulation resistances.

Stage 2

Dismantling the motor

Dismantle the motor as follows:

(a) Remove any coupling or pulley with a "three leg" puller.

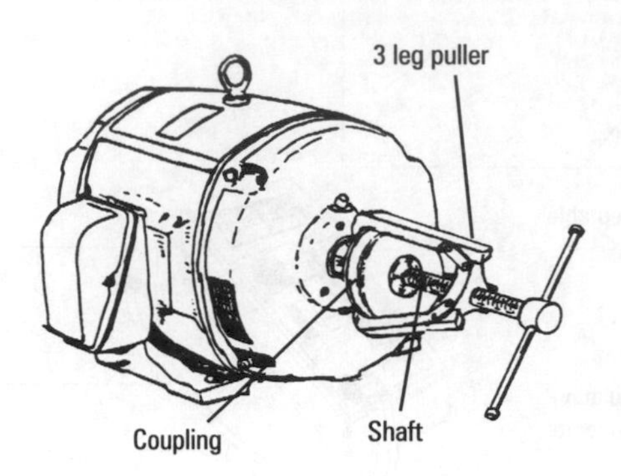

Figure 4.24

(b) Remove the drive end bearing plate from the endshield.

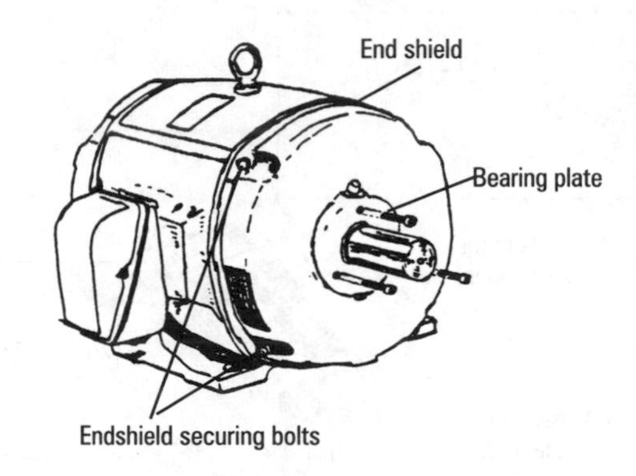

Figure 4.25

(c) Remove fan cover and fan impeller.

(d) Mark the frame and both endshields so they can be replaced in the same position.

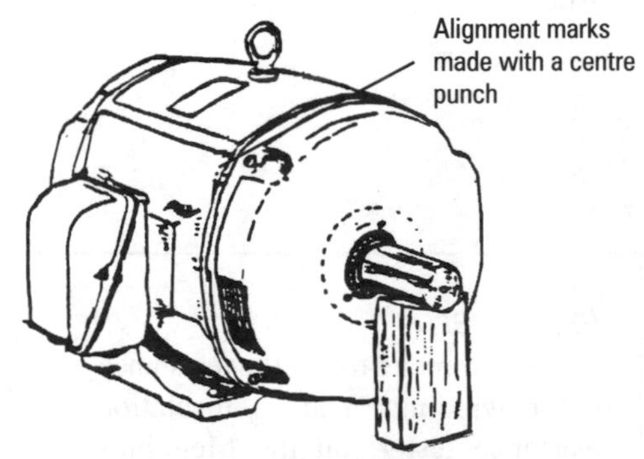

Figure 4.26

(e) Remove the securing bolts from the non-drive end endshield.

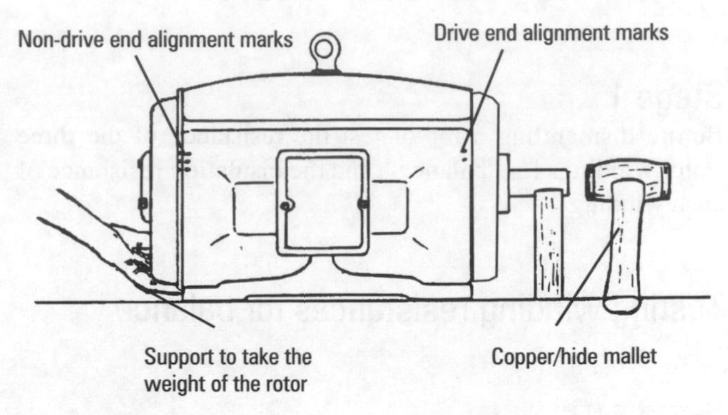

Figure 4.27

(f) Tap the shaft with a mallet to free the non-drive end endshield.

(g) Withdraw the rotor with the endshield.

(h) Unscrew and remove the bearing plate.

(i) Support the endshield and gently tap the rotor shaft to remove the endshield.

(j) Remove the drive end securing bolts

(k) Remove the drive end endshield by tapping gently from the inside.

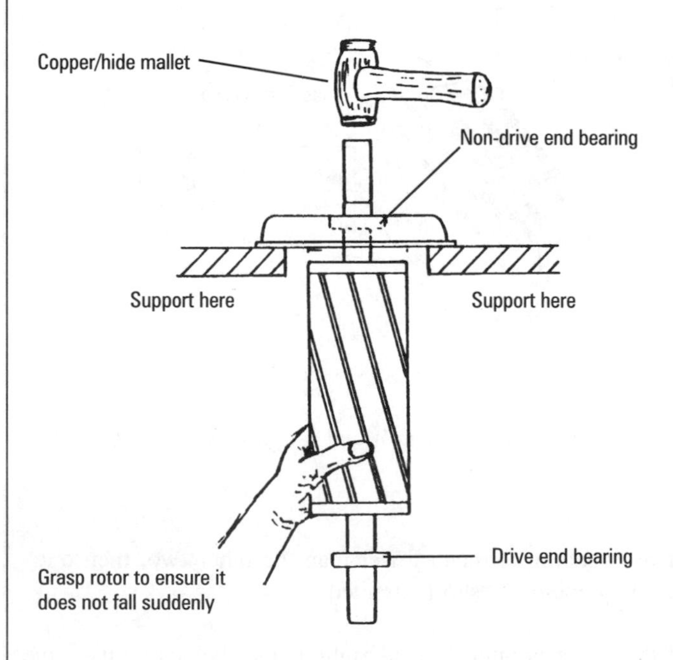

Figure 4.28

Stage 3

Inspection of component parts

Inspect all motor parts for serviceability and cleanliness using check list.

Check:

- the bearings for wear (look for rub marks around the periphery of the rotor)
- that the bearings are adequately greased and that no grease has been thrown out of the bearings into the motor
- for signs of overheating
- that the stator windings are not damaged and are securely wedged into their slots
- that there are no loose rotor bars
- for ingress of dust and that ventilation holes are clear
- that the fan has not been rubbing on the cover and that there are no broken fan blades
- the motor frame for damage (cracks)
- that terminations are electrically and mechanically sound

Stage 4

Reassembly of the motor

The procedure for reassembly is the reverse of dismantling.

Points to note are that:

- all parts are clean
- mating parts are properly aligned
- the rotor turns freely

Test windings for:

- continuity and
- insulation resistance (as in Stage 1) before connecting motor to the supply.

Single-phase motors can be dismantled and reassembled in a very similar way taking care to ensure that the centrifugal switch is not damaged in the process.

Special lifting gear is required on larger motors to withdraw or replace the rotor as shown in Figure 4.29.

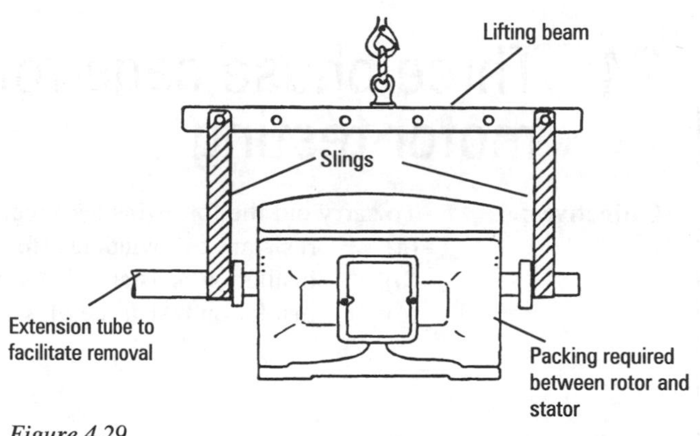

Figure 4.29

Lifting method

1. Take the weight gently and move the rotor sideways until one sling is close to the frame.
2. Place a support block under the shaft and lower the rotor so that it rests in the stator and on the support block.
3. Reposition slings, take the weight, remove the support, and move the rotor sideways.
4. Lower the rotor carefully as in (2).
5. Repeat stages (3) and (2) until the rotor is clear of the stator.
6. Rest the rotor on two support blocks.

24 Three phase cage-rotor induction motor testing

Objectives: To carry out the following tests, before dismantling and after reassembly, and record readings:

(a) resistance of windings (for balance)
(b) insulation resistance between windings
(c) insulation resistance of windings to metal frame (earth)

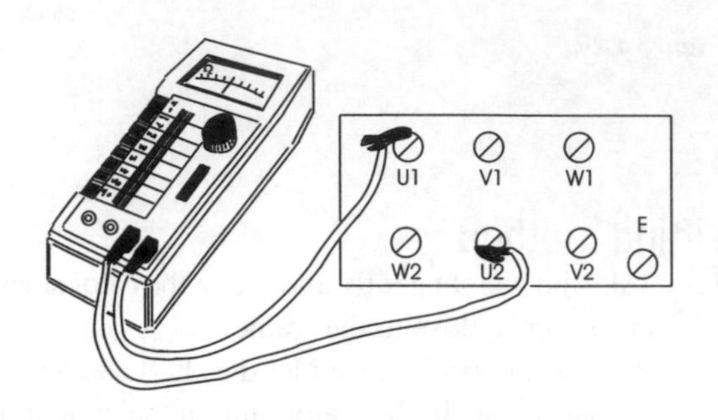

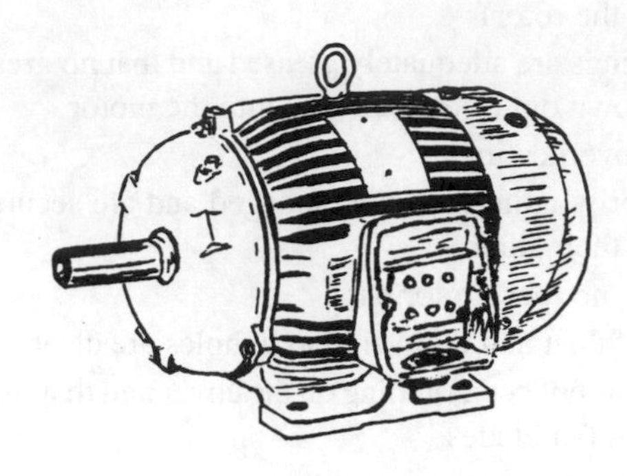

Figure 4.30

Suggested time: 15 minutes

Measurements:

		before	after
(a) winding resistance	U_1 to U_2		
	V_1 to V_2		
	W_1 to W_2		
(b) insulation resistance between windings			
	U and V		
	V and W		
	W and U		
(c) insulation resistance of windings to earth			
	U		
	V		
	W		

Motor required: three phase induction motor any size up to 750 W (1 h.p.)

Test equipment: continuity tester (ohmmeter)
insulation resistance tester (500 V)

Points to be considered:

✓ are the correct measuring instruments being used?

✓ are the measurements acceptable?

25 Three phase cage-rotor induction motor examination

Objectives: To dismantle the machine taking precautions to avoid damaging the machine. (Ensure both end covers and the frame are marked before dismantling.)

To sketch and label the component parts in the box provided below.

To inspect all motor parts for serviceability and cleanliness.

To reassemble machine and check that the rotor turns freely.

Suggested time: 1.75 hours

Motor required: three phase induction motor any size up to 750 W (1 h.p.)

Tools required:
three leg puller
suitably sized spanners – open-ended, ring or socket, and nut spanners
copper/hide mallet (soft mallet)
centre punch
drift (non-ferrous)
suitably sized screwdrivers

Points to be considered:

✓ has the motor been proof marked?

✓ have all component parts been identified correctly?

✓ are mating parts aligned correctly?

✓ does the rotor turn freely?

Examination of a single-phase capacitor start motor

Stage 1

Testing continuity of windings

Before dismantling a motor test the resistance of the start and run windings with an ohmmeter.

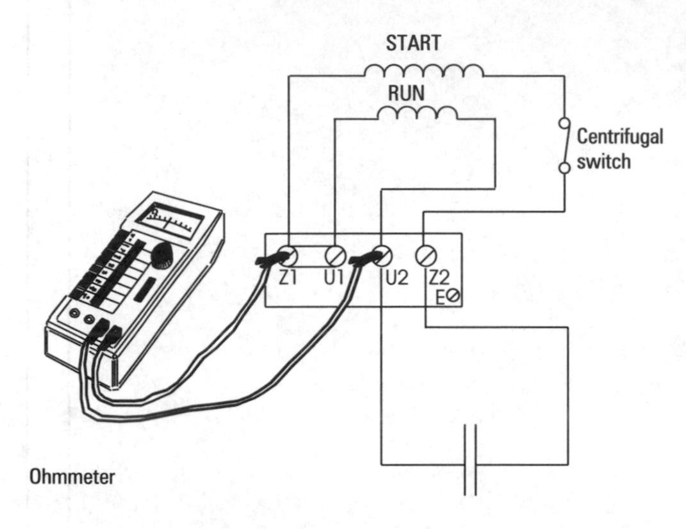

Figure 4.31

Measure between

Z_1 and Z_2

U_1 and U_2

The supply connections are U_1 and U_2.

The resistance of Z_1 to Z_2 should be greater.

If there is no reading across Z_1 and Z_2 this may be due to a faulty centrifugal switch and it should be examined when the motor is dismantled.

Testing insulation resistance

This test measures the resistance between the windings and between each winding and the motor earth connections. The instrument measures (MΩ) Megohms and uses a voltage of 500 V so care must be taken when carrying out this test.

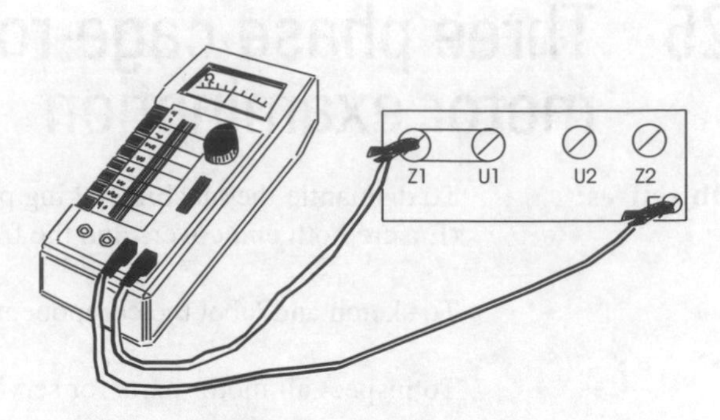

Figure 4.32

As with three phase motors a good insulation resistance value is indicated by a high resistance reading of between 100 MΩ and infinity ∞, and a poor value is indicated by a low resistance reading (a reading of 0.5 MΩ and below is an indication of very poor insulation resistance.)

Stage 2

Dismantling the motor

Carry out the procedure as described for the three phase motor but extra care must be taken when removing the end shield from the non-drive end as the centrifugal switch mechanism may become damaged.

Stage 3

Inspect the motor parts and work through the check list used for the three phase motor but include the switch mechanism of the centrifugal switch.

Stage 4

Reassemble taking care to correctly align the centrifugal switch assembly.

26 Single phase capacitor-start induction motor testing

Objectives: To carry out the following tests and record readings.

(a) resistance of (i) start winding
(ii) run winding

(b) insulation resistance between (i) start and run windings
(ii) start winding to earth
(iii) run winding to earth

After reassembly
(c) to carry out tests as above and
(d) to check and test capacitor

SAFETY NOTE: take care not to exceed the working voltage of the capacitor

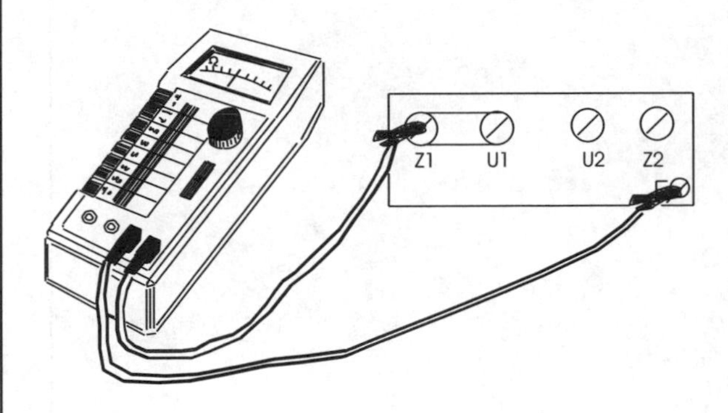

Figure 4.33

Suggested time: 15 minutes

Measurements: before after

(a) (i) start winding resistance ____________________
(ii) run winding resistance ____________________

(b) insulation resistance between
(i) start and run windings ____________________
(ii) start windings to earth ____________________
(iii) run windings to earth ____________________

Motor required: single phase capacitor start motor any size up to 750 W (1 h.p.)

Test equipment: insulation resistance tester (500 V)
continuity tester (ohmmeter)
suitably sized discharge resistor

Points to be considered:

✓ are the correct measuring instruments being used?

✓ are the measurements acceptable?

27 Single phase capacitor-start induction motor examination

Objectives: To dismantle the motor taking precautions to avoid damaging the motor.

To sketch and label the component parts for serviceability and cleanliness.

To check operation of centrifugal switch mechanism.

To reassemble motor and check that the rotor turns freely.

Suggested time: 1.75 hours

Motor required: single phase capacitor start motor any size up to 750 W (1 h.p.)

Tools required: suitably sized spanners – open-ended, ring or socket, and nut spanners
copper/hide mallet (soft mallet)
centre punch
drift (non-ferrous)
suitably sized screwdrivers

Points to be considered:

✓ has the motor been proof marked?

✓ have all components parts been identified correctly?

✓ are mating parts aligned correctly?

✓ does the rotor turn freely?

Examination of a universal (series) motor

Stage 1
Testing continuity of windings

Before dismantling, and after reassembling the motor, test for continuity through the field windings and armature.

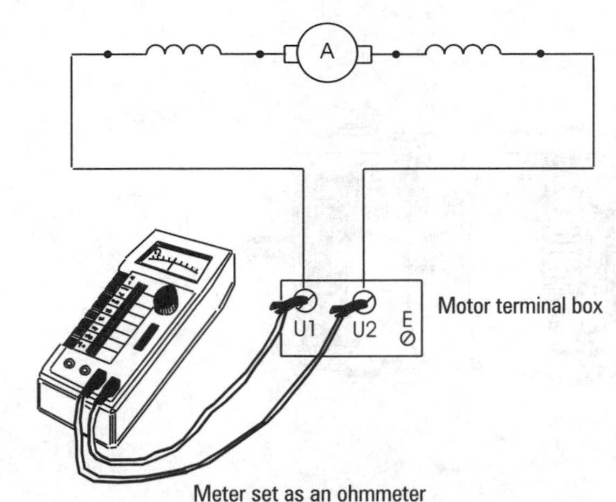

Figure 4.34

If there is continuity note the resistance reading. The shaft of the motor should now be rotated so that the brushes rest on the next segments of commutator. Continue to do this noting the reading at each stage until each pair of segments have been tested. A variation in the readings, or no reading at all, may be due to bad brush contact or a carbon film on the commutator.

Testing insulation resistance

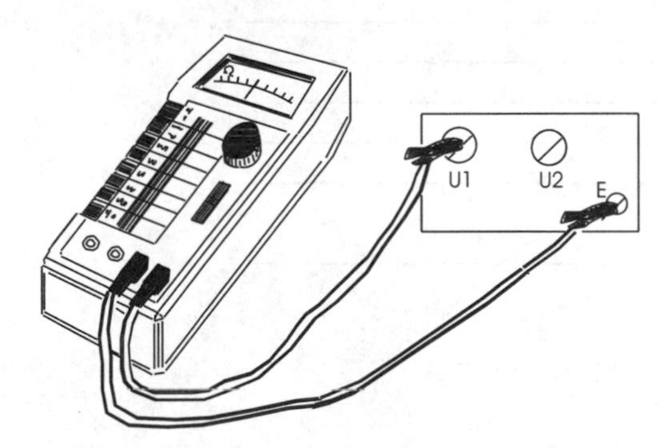

Figure 4.35

A good insulation resistance value is indicated by a high resistance reading of between 100 MΩ and infinity ∞, and a poor value is indicated by a low resistance reading (a reading of 0.5 MΩ and below is an indication of very poor insulation resistance.)

Stage 2
Dismantling the motor

Make sure the end shields are marked for position on the frame. Before the motor is dismantled the brushes must be removed. The end shields can now be unbolted from the frame. Remember that there are connections between the field winding and the end shield containing the brush holders.

Stage 3
Inspection of component parts

Inspect all motor parts for serviceability and cleanliness using check list.

Check list:

General:

- bearings for wear (look for rub marks around the periphery of the armature)
- that bearings are adequately greased and that no grease has been thrown out of the bearings into the motor
- for signs of overheating
- that the field windings are not damaged and are secure
- that there are no damaged armature windings
- for ingress of dust and that ventilation holes are clear
- that the fan has not been rubbing on the cover and that there are no broken fan blades
- the motor frame for damage (cracks)
- that terminations are electrically and mechanically sound

Brushes:

- for wear
- slide freely in their holders
- bed correctly on the commutator surface
- for spring tension
- brush lead connections are mechanically sound

Commutator:

- for wear (any signs of surface grooving?)
- for discolouration (any signs of arcing between the segments?)
- insulation is below the surface of the copper segments (adequately undercut)

Stage 4

Reassemble the motor ensuring that the commutator is correctly aligned with the brushgear and the end shields are correctly aligned with the frame.

Stage 5

Test for continuity and insulation resistance as in Stage 1.

28 Universal (a.c./d.c.) motor testing

Objectives: To carry out the following tests, before dismantling and after reassembly, and record the readings:

(a) continuity of windings
(b) insulation resistance of windings to metal frame (earth).

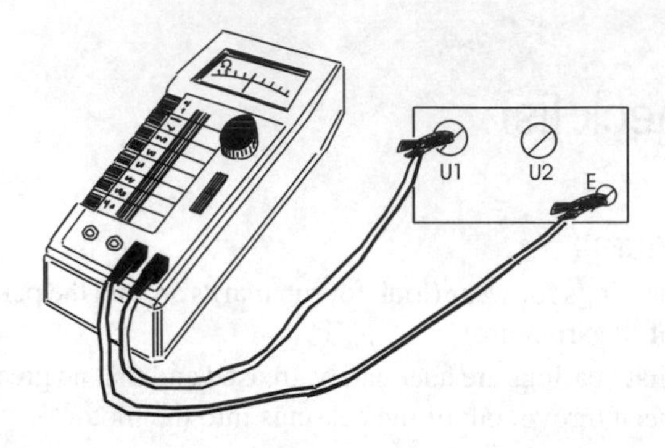

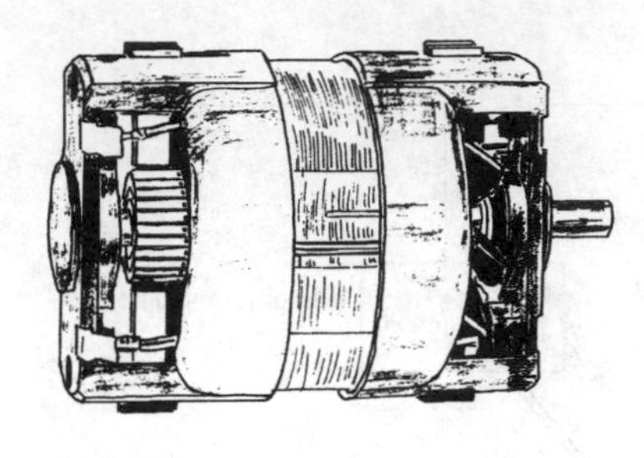

Figure 4.36

Suggested time: 15 minutes

Measurements:

	before	after
(a) winding continuity		
(b) insulation resistance		

Motor required: universal motor any size up to 750 W (1 h.p.)

Test equipment: insulation resistance tester (500 V)
continuity tester (ohmmeter)

Points to be considered:

✓ are the correct measuring instruments being used?

✓ are the readings acceptable?

29 Universal (a.c./d.c.) motor examination

Objectives: To dismantle the machine taking precautions to avoid damaging the machine. (Ensure both end covers and the frame are marked before dismantling.)

To sketch and label the component parts in the box provided below.

To inspect all motor parts for serviceability and cleanliness,

To reassemble machine and check that the armature turns freely.

Suggested time: 1.75 hours

Motor required: universal motor any size up to 750 W (1 h.p.)

Tools required:
suitably sized spanners – open-ended, ring or socket, and nut spanners
copper/hide mallet (soft mallet)
centre punch
drift (non-ferrous)
suitably sized screwdrivers

Points to be considered:

✓ has the motor been proof marked?

✓ have all components parts been identified correctly?

✓ are mating parts aligned correctly?

✓ have the brushes been correctly fitted?

✓ does the armature turn freely?

30 Forward and reverse direct-on-line starter

Objectives: To wire up the starter circuit as shown, ensure that all enclosure covers are replaced and that all exposed metalwork is earthed and bonded.
To carry out appropriate inspection and tests.
To switch on the supply and test run the motor, ensuring no danger exists from moving parts.

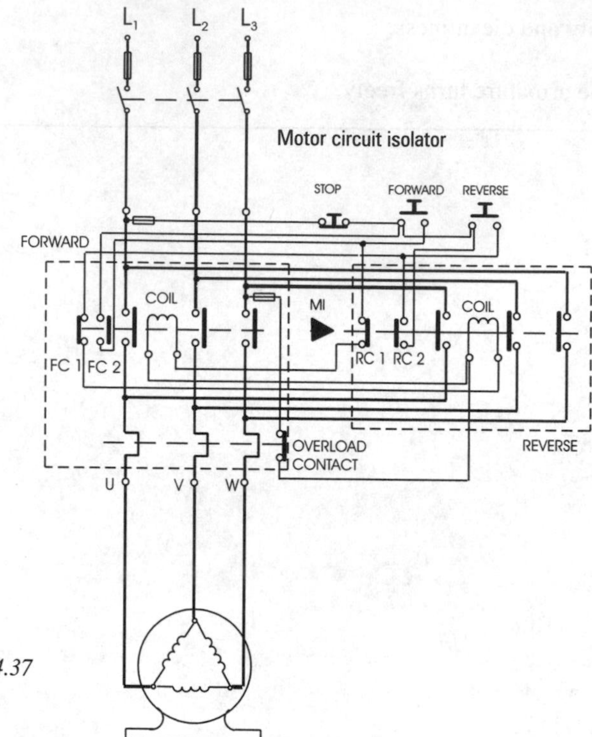

Figure 4.37

Legend (key) to diagram

FC 1 and
RC 1 Electrical interlocks
FC 2 and
RC 2 Hold-on contacts
M1 Mechanical interlock

Control circuit

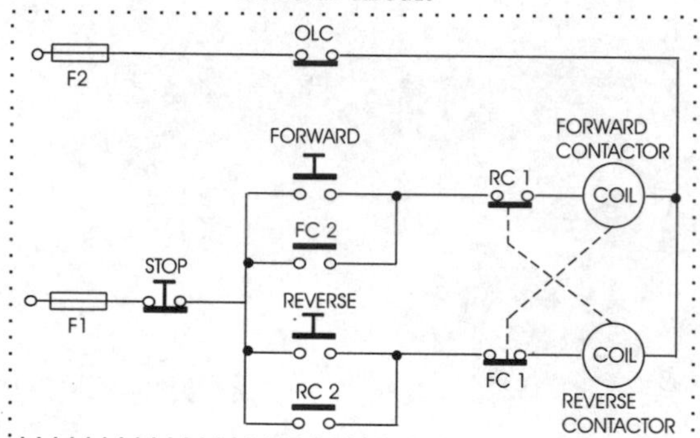

Suggested time: 2 hours

Equipment: forward and reverse D-O-L starter with 400 V coils
three-phase cage rotor motor
insulation resistance tester

Points to be considered:

✓ has the wiring been carried out correctly?

✓ are all terminations secure?

✓ are all terminations shielded from direct contact?

✓ have all enclosures been securely fastened?

✓ has the wiring been carried out in a safe manner?

✓ have the objectives been achieved?

WARNING!
This exercise should be carried out in accordance with the Electricity at Work Regulations 1989 and all live conductors should be shielded and insulated from touch. All test probes should meet the requirements of the Health and Safety Executive Guidance Note GS38.

5

Commissioning and Testing Procedures

On completion of this chapter you should be able to:

- carry out commissioning in accordance with BS 7671 for an a.c. motor
- carry out commissioning in accordance with BS 7671 for a radial circuit
- carry out commissioning in accordance with BS 7671 for a lighting circuit
- carry out tests and take readings using appropriate instruments as identified with BS 7671
- test earth leakage devices

Introduction

Inspection

The inspection of an installation should be an ongoing process while the wiring systems are being installed. This can often save time in the future when much of the wiring is concealed. There are certain things that must be inspected at different stages in the installation process. For example all conduit runs must be complete before any cables are installed and to confirm they are complete and all connections are tight, an inspection must be carried out.

An inspection can often reveal odd things that have been overlooked and could be dangerous if left. The equipotential bonding conductor that was installed but couldn't be connected at the time because the room was not complete is an example. Without an inspection this could be missed and some item left unbonded. Inspection does not stop at just looking at completed work, it may involve removing accessory plates and looking inside, checking connections are tight, all mechanical protection is in place, cord grips are used correctly....

It would be helpful to refer to the inspection checklist which can be found in IEE Guidance Notes 3 "Inspection and Testing".

The actual tests and the procedures for carrying them out will be found in the book "Stage 1 Design" and it will be helpful to you to have this book available for reference before carrying out the exercises. The following notes cover some of the background information to testing.

To carry out all of the tests identified in BS 7671:1992 a variety of instruments is required. Each instrument requires a voltage to carry out the test. The voltages vary from extra low to high. In every case care needs to be taken to ensure that the tests are carried out safely.

In all there are eleven different tests identified in BS 7671 but it is seldom that all of these will apply to a single installation. The tests should be carried out in the sequence shown to ensure that all tests are completed and any faults are identified.

You will find a note of the tests in Part 7 in BS 7671:1992, Section 713.

The list of tests can be divided into two, those that are carried out before the supply is connected and those that require the supply for the test.

Those that are carried out before the supply is connected:

- continuity of protective conductors
- continuity of ring final circuit conductors
- insulation resistance
- site applied insulation
- protection by separation of circuits
- protection against direct contact, by a barrier or enclosure provided during erection
- insulation of non-conducting floors and walls
- polarity
- earth electrode resistance

Those that require the supply to be connected before they can be carried out:

- polarity (confirmation)
- earth fault loop impedance
- residual current operated devices

In addition to these it will be necessary to carry out a prospective short circuit current test at the intake to, and distribution boards of, the premises.

Safety when testing

Although many of the installations that have to be tested are new and the supply may not be connected, others will have been energised for years. In these cases the section of installation under test must be isolated for the first group of tests. Even after an inspection has been carried out it is still possible in some cases for voltages to exist even though everything appears to be dead.

To ensure there are no unexpected supplies available when tests are to be carried out, an isolation procedure should be adopted similar to that shown in Figure 2.5. A voltage indicator similar to one of those shown in Figures 5.1 and 5.2 are usually most suitable for this test.

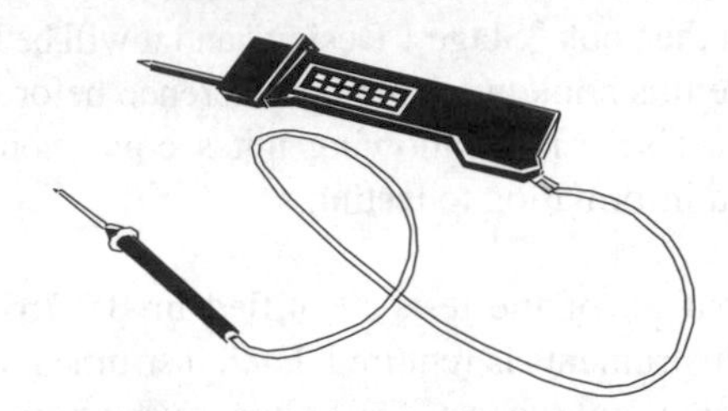

Figure 5.1

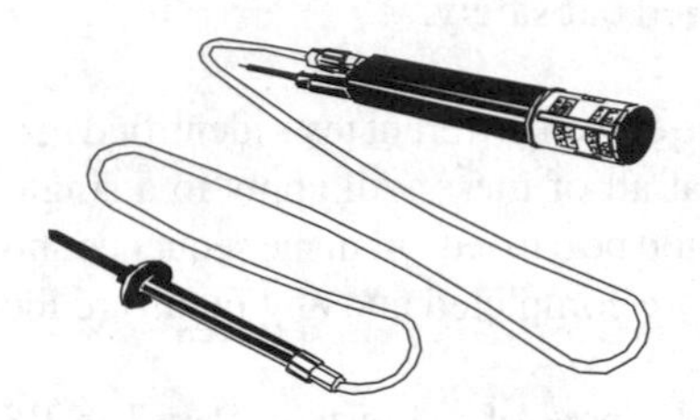

Figure 5.2 *Examples of voltage indicators*

The Health and Safety Executive have produced Guidance Note GS 38 covering "electrical test equipment for use by electricians". In this they have identified that many electricians have received burns due to arcing or flashover resulting from the use of unsatisfactory test probes. To overcome this they have recommended that test probes and leads should have shrouded connectors to the test instrument, robust flexible well insulated leads, HBC fuses or other current limitation, finger barriers to guard against inadvertent hand contact and have a maximum of exposed metal tip of 2mm.

In general the person carrying out the tests should be competent. BS 7671:1992 does not give a definition of competence but it does state that competent persons will (Appendix 6 Introduction, iv),

> "have a sound knowledge and experience relevant to the nature of the work undertaken and to the technical standards set down in this British Standard, be fully versed in the inspection and testing procedures contained in this Standard and employ adequate testing equipment".

The Health and Safety Executive have defined a "competent" person as,

> "A person with enough practical and theoretical knowledge and actual experience to carry out a practical task safely and effectively. The person should have the necessary ability in the particular operation of the type of plant and equipment with which he or she is concerned, an understanding of relevant statutory requirements and an appreciation of the hazards involved. That person should also be able to recognise the need for specialist advice assistance when necessary and to assess the importance of the results of examinations and tests in the light of their purpose."

Some indication can be gained from these as to the experience and knowledge required of the person qualified to carry out the tests.

Figure 5.3 *Competent?*

Frequency of tests

An inspection and the tests should be carried out on all new works. In addition to this they need to be carried out when there are alterations made or where there is a significant change in the loading or use of the premises. At periodic intervals all installations should be checked. The exact amount of time depends on the building and its use. The intervals between inspections are the maximum and need to be reviewed regularly. Complete Table 5.1 using Table 4A in IEE Guidance Notes 3 Inspection and Testing:

Table 5.1

Type of building	Maximum period between inspection and testing
Cinemas	
Theatres	
Caravan parks	
Petrol filling stations	
Domestic	
Places of public entertainment	
Agricultural and horticultural	
Construction site installations	

If any of the tests carried out show a failure to comply, that particular test, and any preceding test where the results may have been influenced by the fault indicated by the failure, will need to be repeated after the fault has been corrected.

Test instruments

To ensure the readings obtained in tests are as accurate as possible the test instruments should be continually checked. In some cases this is a case of zeroing the scale with loads shorted out. In others it may be a case of carrying out battery checks. It is always advisable to note which instrument has been used to carry out particular tests. In this way should an instrument later be found to be faulty an immediate check can be made on what the instrument has been used for.

It is important that readings are accurately transferred from the test equipment and recorded. Where multi range instruments are used care has to be taken to ensure that the correct scale and range is read. Instruments with digital displays have, in some cases, made the readings much easier to record. Even these can be misinterpreted as they often use "floating" decimal points and automatic scaling. In these cases care must always be taken to read the display carefully to determine exactly what the reading is.

It is important that correct equipment is used for each test otherwise the results may be misleading.

Continuity of protective conductors – we shall assume that these are copper conductors either as part of a cable, such as p.v.c. twin and cpc, or a separate single insulated cable. The resistance of these conductors can be very low especially where they have a large cross sectional area. Resistances of less than 1 Ω may need to be recorded so a meter capable of this should be used. As this is is a continuity test and it is checking the cable is continuous throughout its length, a high voltage is not required. Voltages of between 6 and 9 volts are typical.

If the circuit protective conductor is made of steel conduit then a separate high current test may need to be used.

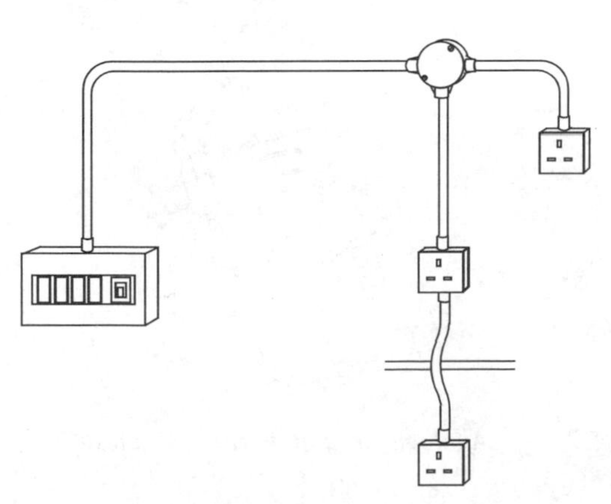

Figure 5.4 *Steel conduit*

Continuity of ring final circuits – is another continuity test that will be recording low resistance values. Where the ring circuit being tested is short the resistance may be below 0.01 Ω.

Insulation resistance tests – unlike the two above are measuring the insulation resistance not the conductor resistance. It is important to ensure that the conductors are not coming together at any part of their length. It is also important to test that the insulation is not likely to break down under pressure. The way of applying this pressure test is to apply a high voltage across the insulation. The test voltage for installations supplied with 230 V or 400 V is 500 V d.c. at a current of 1 mA. As it is insulation that is being measured the results should be in the millions of ohms range.

Polarity – is a test that checks on the correct wiring of circuits and no value is obtained that should be recorded. A continuity meter may be used for this test but often a bell or buzzer is used from a low voltage battery supply.

Earth fault loop impedance tests – are carried out after the supply has been connected. The test instrument injects a current of up to 25 A through the earth fault path and records the resultant value in ohms.

Earth electrode resistance – this is a specialist test which may require special test equipment. However, if the earth electrode to be tested is part of a TT system, an earth fault loop impedance tester may be used.

Operation of residual current devices – requires equipment especially designed for the test. The test equipment injects small currents through the RCD and checks if it trips or not. The instrument will display the tripping time in milliseconds.

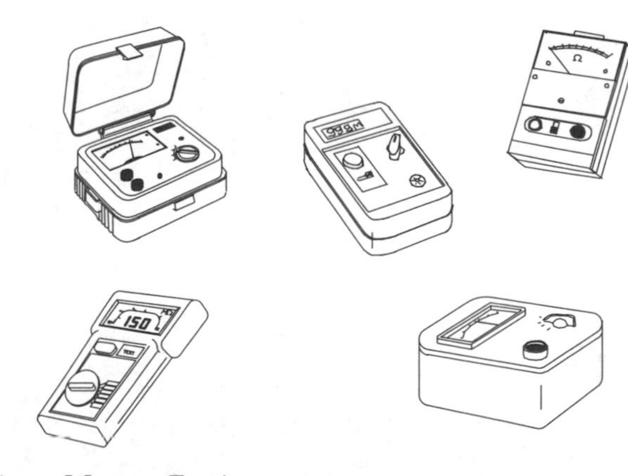

Figure 5.5 *Test instruments*

Question

State the instruments required, and their ranges, for the following tests:

(a) continuity of protective conductors
(b) continuity of ring final circuits
(c) insulation resistance
(d) polarity
(e) earth fault loop impedance

Answer

The instruments required for the tests are as follows:

(a) Continuity of protective conductors – an ohmmeter with a range starting at less than 1 ohm.

(b) Continuity of ring final circuits – an ohmmeter with a range starting at less than 0.1 ohm.

(c) Insulation resistance – an ohmmeter that uses a test voltage of 500 V d.c. at a current of 1 mA and measures resistance values up to 1 MΩ

(d) Polarity – an instrument similar to that used for continuity testing or a bell set that will operate on resistances up to 1 ohm.

(e) Earth fault loop impedance – this instrument must be calibrated in ohms but works on currents up to 25 A. The resistance range must be capable of readings down to 0.1 of an ohm.

Try this

Explain what safety precautions must be taken before any tests are carried out?

Tip:

Section 10 of the On Site Guide may help when looking at safety.

Insulation resistance testing

All testing should be carried out in such a way that safety is always given a top priority. When insulation resistance tests are to be carried out it must be remembered that a voltage of 500 V d.c. is used. Although this is not a lethal voltage it can create serious situations. For example someone standing on steps putting up a luminaire could receive a shock from the test voltage and fall.

Insulation resistance tests are often carried out on extensions to installations or are part of a periodic inspection and test. In these situations live supplies are present. Before an insulation resistance test can be carried out checks should be made to ensure no live conductors are accessible. An isolation test procedure should be adopted before any insulation resistance test instrument is connected.

Many insulation resistance test instruments now use batteries as their source of power. If the supply of these is low, incorrect readings will be given. Carry out a battery check before any recordings are made.

After a time test instruments become inaccurate and it is important that calibration is carried out at regular intervals.

Over the years many insulation resistance testers have been developed but the specifications that were used previously do not necessarily meet the accepted requirements now. It is always important to check that the instruments used are "up-to-date" and the readings they give are valid.

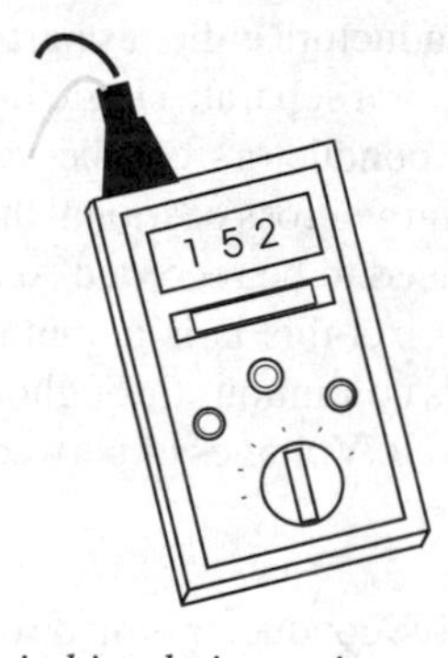

Figure 5.6 Digital insulation resistance tester

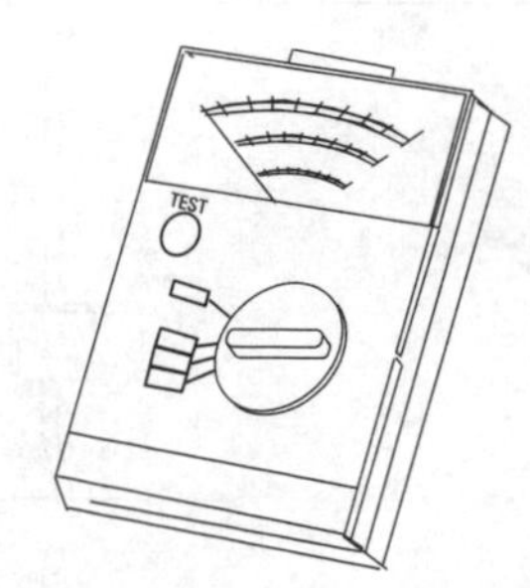

Figure 5.7 Analogue insulation resistance tester

So that no confusion can exist as to what instrument has been used for particular tests the serial number of the equipment should always be recorded on the results sheet.

Before any tests are carried out the condition of the test leads should be checked for damage. Tests should be carried out with the leads open circuit and short circuited to ensure the correct readings are obtainable.

An insulation resistance test is carried out to confirm there is no connection between the live conductors or the live conductors and earth. If equipment is in the circuit and connected between phase and neutral or between phases the test voltage will also be put across this. Apart from the fact false readings will be given, serious damage may be caused to the equipment. Before testing between phase and neutral ALL equipment, lamps, electronic devices and portable appliances must be disconnected. Electronic dimmer switches must be shorted out so that they are not damaged and all of the circuit can still be tested.

When all of this has been carried out testing can be considered.

Question

With the aid of diagrams explain how to carry out an insulation resistance test on a final circuit from a single-phase distribution board. It is known that at least one circuit contains an electronic dimmer switch.

Answer

The first consideration when carrying out an insulation resistance test from a distribution board must be safety. Tests must be carried out to ensure that the final circuit to be tested is "dead" and there are no exposed live conductors that could be touched when carrying out the test. Once it has been confirmed the circuit is safe to work on an inspection of the circuit should be carried out. All lamps should have been removed and switches put in the "on" position. Fluorescent luminaires must have been disconnected or unplugged. Dimmer switches should have been shorted out and a link put across two-way switches. When it is clear that all of this is complete the test can be carried out.

First check the meter for battery condition and see that the leads and connections are not damaged. Short the leads together and test the instrument. If everything is working correctly, connect one lead to the circuit neutral and the other to the phase conductor (Figure 5.8).

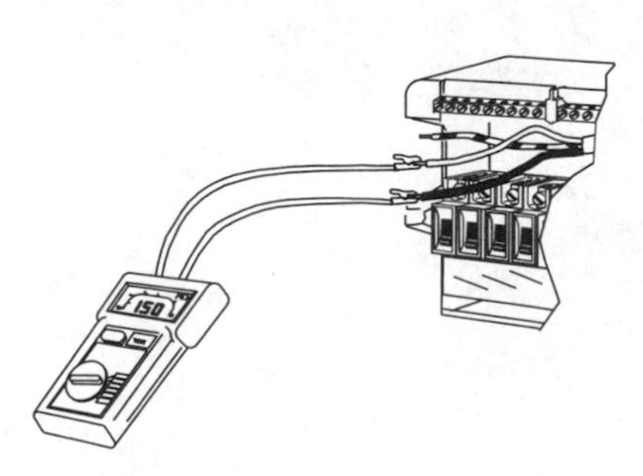

Figure 5.8

When the reading is taken the resistance should exceed 0.5 MΩ. With a short lead connect the phase and neutral together and test between this link and the circuit protective conductor.

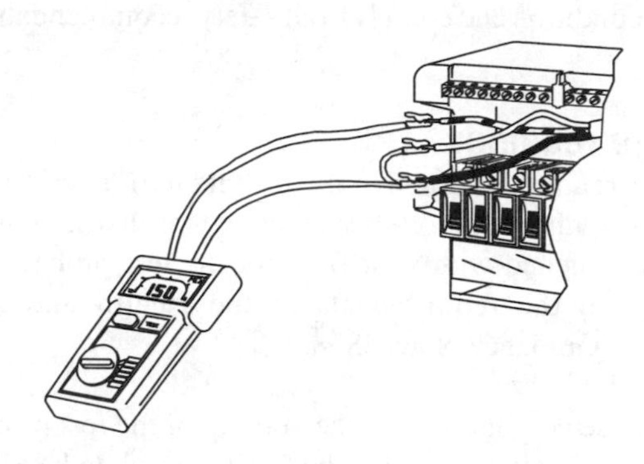

Figure 5.9

This reading should also be greater than 0.5 MΩ. When the readings are complete the circuit should be restored to its original condition, i.e. links removed, lamp replaced and equipment reconnected.

Try this

Explain how you would carry out an insulation resistance test on a three-phase delta connected induction motor.

Tip:

The details given in Section 10.3.3 of the IEE On Site Guide should be of help but need modification to apply to the three-phase motor.

Voltage and current readings

It is sometimes necessary to take voltage and current readings on circuits when the supply is connected. These should always be carried out with care and in such a way that the possibility of getting a shock is negligible. To do this the equipment must be in good condition and comply to all safety recommendations.

Voltage readings

Voltage readings should only be taken if everything is shrouded with insulation to such an extent that nothing live could be touched with your fingers. The test probes should conform to the requirements of the Health and Safety Executive Guidance Note GS38.

Under these conditions only the very tip of the test probe can touch the live conductors and the flexible insulated leads with shrouded ends ensure the voltages to the test instrument are safe. Under these conditions voltages can be taken, with care, with very little or no risk of shock.

Current readings

To use a conventional ammeter to take current readings, the circuit has to be broken and the ammeter connected in series with the load. Using a clamp-on type ammeter no circuit connections have to be connected or disconnected. The only requirement is that it is possible to get at a single circuit conductor. This conductor must be insulated. The ammeter can be clamped around this and the reading taken in amperes. This instrument should only be used on single insulated conductors where there are no exposed live parts.

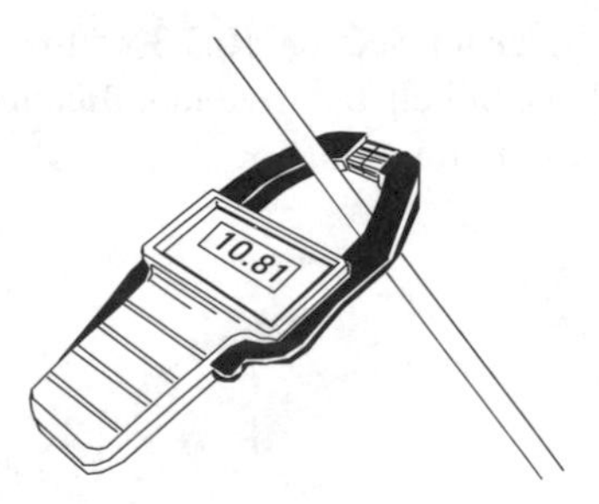

Figure 5.10 Clamp-on ammeter

Question

The working voltage and current of each phase is to be taken at a three-phase induction motor.

A three-phase cage induction motor is installed at the end of a long cable run. Checks have to be carried out to ensure the calculations for voltage drop and load current conform to the actual equipment when in use. Describing all necessary safety precautions explain how the

(a) voltage can be measured
(b) current in each phase can be determined when the motor is working on full load

Answer

(a) When measuring the voltage at an a.c. three-phase motor care must be taken first to ensure there is no chance of injury due to the mechanical movement of the motor drive mechanism or the load. Next the correct instrument must be selected. This would be a meter set to a.c. voltage and capable of readings up to and including 400 V. The leads and connections should be checked to ensure the insulation is not damaged. The test prods should be of the appropriate type.
With the motor isolated the terminal cover should be removed. If necessary conductors should be adjusted so that a clear path and sight is available to each connection. It may be advisable to try the test probes on each terminal while the motor is still isolated. If there is any chance of personal contact with terminals while carrying out the tests then extra insulation may need to be put in place. When satisfied that the tests can be carried out safely the motor should be started up. The voltages between each of the phases should be measured, recorded and compared with the design calculations. The terminal cover should be replaced before leaving.

(b) As the load current of the motor will be the same throughout the length of cable run, the current does not have to be measured at the motor terminals. A clamp-on ammeter would be used as this does not require breaking into the circuit. Any point in the cable run where individual phase conductors can be made accessible will be suitable. Making sure there is no exposed live parts the motor should be run up on load. The ammeter should be clamped round each individual conductor in turn and the three currents recorded. These should be compared with the calculated design current. All enclosures should be returned to their safe condition before leaving.

Try this

Explain why it is necessary for connections to be shielded to IP2 when voltage tests are to be carried out.

Tip:
Look at a copy of the IP code and see what restrictions IP2 offers.

Remember

Table 5.2 shows the instruments used for each of the tests.

Tests	Test Equipment
Continuity of protective conductors	copper conductor – low resistance ohmmeter (milliohmmeter)
Continuity of ring final circuit conductors	low resistance ohmmeter (milliohmmeter)
Insulation resistance	high resistance ohmmeter with d.c. test voltages to suit the installation under test, i.e. 250 V, 500 V or 1000 V when loaded with 1 mA. Resistance range needs to be in excess of 1 MΩ.
Site applied insulation	high resistance ohmmeter with 4000 V a.c. test voltage applied for 1 minute
Protection by separation of circuits	Test 1 high resistance ohmmeter with 500 V d.c. test voltage Test 2 high resistance ohmmeter with 4000 V d.c. test voltage
Protection against direct contact, by barrier or enclosure provided during erection	Test probes with IP2X or IPXXB and IP4X, 40 V to 50 V supply and test lamp
Insulation of non-conducting floors and walls	Test 1 high resistance ohmmeter with 500 V d.c. test voltage and electrodes conforming to IEC 364-6-61, 1986 Test 2 test voltage of 2000 V a.c. and measurement of leakage current up to 1 mA.
Polarity	low resistance ohmmeter (milliohmmeter)
Earth fault loop impedance	loop impedance tester
Earth electrode resistance	**either** extra low voltage a.c. supply, voltmeter, ammeter and test electrodes **or** an earth tester and test electrodes **or** in some cases a loop impedance tester may be suitable
Operation of residual current operated devices	RCD tester with suitable ranges

31 Commissioning an a.c. motor

Objective: To carry out the commissioning of a three-phase cage-rotor induction motor in accordance with BS7671.

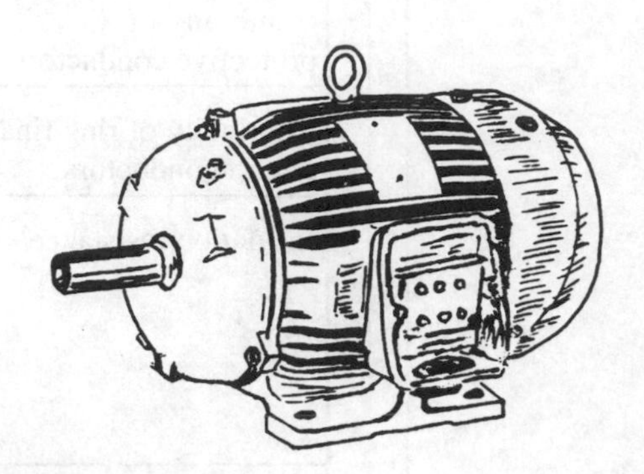

Figure 5.11

Suggested time: 2 hours

Inspection check list:

		Tick box accept	Tick box reject	Comments
Motor circuit:	cable size	❑	❑	
	circuit protection	❑	❑	
	wiring system	❑	❑	
	cpc size	❑	❑	
	isolation device	❑	❑	
	starter type/rating	❑	❑	
	no-volt protection	❑	❑	
	overload protection	❑	❑	
	stop/start control	❑	❑	
Motor:	free to rotate	❑	❑	
	coupling secure/aligned	❑	❑	
	cooling system	❑	❑	
	enclosure undamaged	❑	❑	
	enclosure type	❑	❑	
	installed correctly	❑	❑	
	securely bolted down	❑	❑	
	complies to BS and IP	❑	❑	
	design meets specification	❑	❑	
	terminations correctly connected	❑	❑	
	terminations secure	❑	❑	
	guards secure	❑	❑	

Testing check list

Measurements

Motor circuit:	cpc continuity	❑	❑	
	insulation resistance	❑	❑	
	1. between phases	❑	❑	
	2. between phases and earth	❑	❑	
Motor:	winding resistance balance	❑	❑	
	insulation resistance between windings to earth	❑	❑	
Operating conditions				Comments
	correct direction of rotation	❑	❑	
	noise level	❑	❑	
	vibration level	❑	❑	
	operating temperature	❑	❑	
	operating speed	❑	❑	

Note: the type of installation is left to the tutor's discretion.

Motor required: Three-phase cage-rotor induction motor any size up to 2 kW.

Points to be considered:

✓ have relevant personnel been contacted?

✓ has the motor circuit been safely isolated?

✓ have all safety precautions been taken?

✓ have the inspections been carried out correctly?

✓ are the correct measuring instruments being used?

✓ have the measuring instruments been confirmed as accurate?

✓ are the measurements acceptable?

✓ has a formal record of testing, confirming the safety and integrity of the installation, been prepared?

✓ has the commissioning been carried out correctly?

✓ has the installation been handed over to relevant people with the appropriate information and documentation?

32 Commissioning a radial circuit

Objective: To carry out the commissioning of a radial socket-outlet in accordance with BS7671.

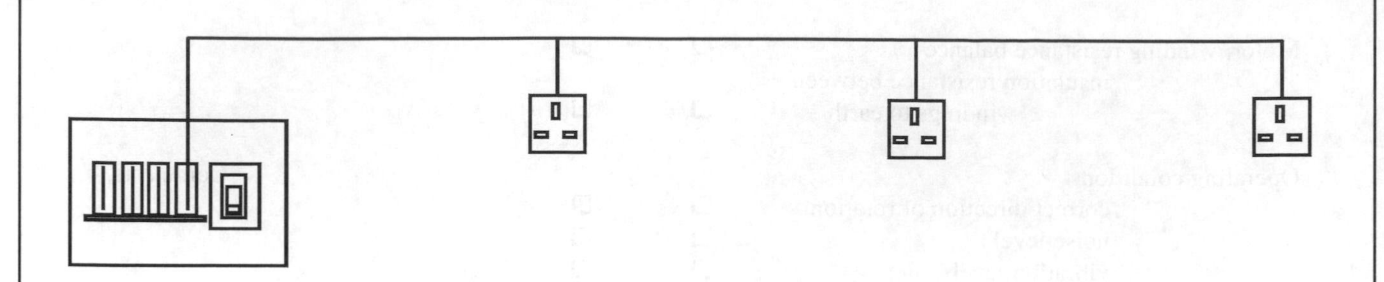

Figure 5.12

Suggested time: 1 hour

Inspection check list:

	Tick box accept	Tick box reject	Comments
cable size	❑	❑	
cpc size	❑	❑	
circuit protection	❑	❑	
wiring system	❑	❑	
isolation device	❑	❑	
circuit designed to specification	❑	❑	
circuit installation	❑	❑	
socket outlet type	❑	❑	
terminations secure	❑	❑	
terminations correctly connected	❑	❑	
accessories secure/undamaged	❑	❑	

Testing check list

	accept	reject	Measurements
cpc continuity	❑	❑	
insulation resistance			
1. P to N	❑	❑	
2. P and N to E	❑	❑	
polarity	❑	❑	

	accept	reject	Comments
circuit functions correctly	❑	❑	

Note: wiring system is left to the tutor's discretion.

Points to be considered:

✓ have relevant personnel been contacted?

✓ has the motor circuit been safely isolated?

✓ have all safety precautions been taken?

✓ have the inspections been carried out correctly?

✓ are the correct measuring instruments being used?

✓ have the measuring instruments been confirmed as accurate?

✓ are the measurements acceptable?

✓ has a formal record of testing, confirming the safety and integrity of the installation, been prepared?

✓ has the commissioning been carried out correctly?

✓ has the installation been handed over to relevant people with the appropriate information and documentation?

33 Commissioning a lighting circuit

Objective: To carry out the commissioning of a two-way and intermediate lighting circuit in accordance with BS7671.

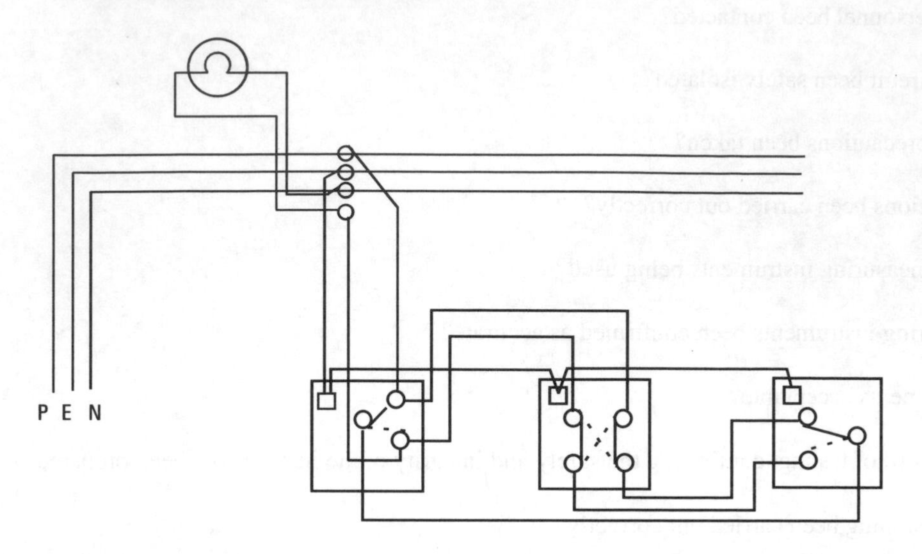

Figure 5.13

Suggested time: 1 hour

Inspection check list:

	Tick box		Comments
	accept	reject	
cable size	❑	❑	
cpc size	❑	❑	
circuit protection	❑	❑	
wiring system	❑	❑	
isolation device	❑	❑	
circuit designed to specification	❑	❑	
circuit installation	❑	❑	
luminaire type	❑	❑	
luminaires and accessories secure/undamaged?	❑	❑	
terminations secure	❑	❑	
terminations correctly connected	❑	❑	

Testing check list

			Measurements
cpc continuity	❑	❑	
insulation resistance			
1. P to N	❑	❑	
2. P and N to E	❑	❑	
polarity	❑	❑	

			Comments
circuit operation	❑	❑	

Note: wiring system is left to the tutor's discretion

Points to be considered:

✓ have relevant personnel been contacted?

✓ has the motor circuit been safely isolated?

✓ have all safety precautions been taken?

✓ have the inspections been carried out correctly?

✓ are the correct measuring instruments being used?

✓ have the measuring instruments been confirmed as accurate?

✓ are the measurements acceptable?

✓ has a formal record of testing, confirming the safety and integrity of the installation, been prepared?

✓ has the commissioning been carried out correctly?

✓ has the installation been handed over to relevant people with the appropriate information and documentation?

34 Continuity of protective conductors

Objective: To carry out tests to measure and confirm the continuity of copper protective conductors.

Time: related to installation or circuit being tested.

Method:

The object of the tests is to ensure that the protective conductors are continuous throughout their length with no high resistance connections.

These tests should be started at the consumer's main earth terminal and then taken out to the different parts of the installation. On small installations a long lead can be taken from the meter. On larger installations the phase or neutral conductor may be used as a return to the meter. In either case the resistance of the return conductor must be allowed for when recording the resistance of the protective conductor.

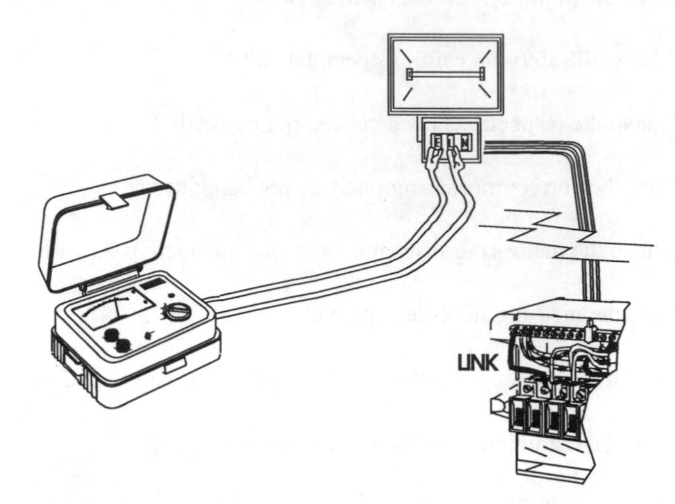

Figure 5.14

Equipment: 1 continuity tester with ohms range
1 real or simulated ring final circuits for testing

Test equipment

Name:
Model:
Serial no.:

Results:

Complete the relevant section of the "Record of Test Results" at the end of this chapter.

Points to be considered:

✓ has the correct instrument been used?

✓ has the serial number of the instrument been noted?

✓ are the batteries in the test equipment up to an acceptable level?

✓ should the instrument be zeroed?

✓ has the instrument been tested open circuit before use?

✓ has the instrument been tested short circuited before use?

✓ has the resistance of the test leads been allowed for?

✓ are the resistances of the circuit protective conductors acceptable?

Questions:

1. On some circuits the circuit protective conductor has a greater resistance than the phase or neutral . Explain why this is so.

2. Describe how this test can be carried out using the phase or neutral as a return conductor to the meter.

WARNING!

This exercise should be carried out in accordance with the Electricity at Work Regulations 1989 and all live conductors should be shielded and insulated from touch. All test probes should meet the requirements of the Health and Safety Executive Guidance Note GS38.

35 Verification of ring final circuit continuity test

Objective: To carry out tests to verify the condition of a ring final circuit

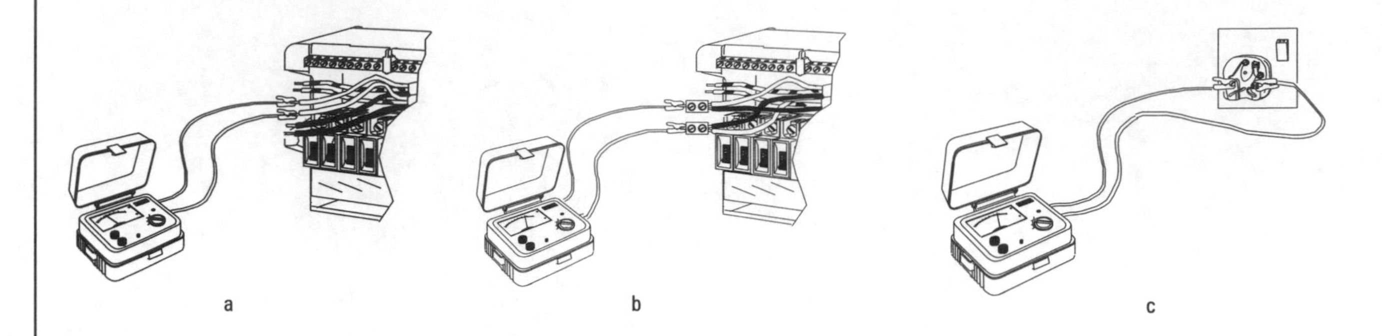

Figure 5.15

Time: related to installation being tested

Method:

To carry out this test an ohmmeter with a range capable of measuring accurately values of less than 1 Ω is required.

The phase conductors are disconnected from the protective device and the ohmmeter is connected across them to confirm there is a circuit. This reading should be noted (Figure 5.15a). The neutral conductors are then disconnected and tested in the same way. This in itself is not sufficient to confirm that the circuit is connected as a ring but if there is no reading at this point then there is no ring circuit continuity. Assuming there is a reading the neutral conductor from one end of the ring is then connected to the phase conductor of the other end of the ring. this is repeated for the other phase and neutral. An accurate resistance reading should now be taken across the two connections and the reading noted. This should be approximately half of the previous reading.

With the circuit still connected as in Figure 5.15b the meter should now be connected across the phase and neutral at each socket outlet. The resistance should be approximately the same as that across the two connections at the board (Figure 5.15c).

Equipment: 1 milliohmmeter of other suitable ohmmeter
1 real or simulated ring final circuit

Test equipment:

Name:
Model:
Serial no.:

Results:

Phase to phase	Ω
Neutral to neutral	Ω
cpc to cpc	Ω
across connectors	Ω

Points to be considered:

✓ has the correct instrument been used?

✓ has the serial number of the instrument been noted?

✓ are the batteries in the test equipment up to an acceptable level?

✓ should the instrument be zeroed?

✓ are the test leads in good condition?

✓ is the circuit isolated from the supply?

✓ are the values of resistance such that the ring final circuit is proved?

Questions:

1. Give examples of TWO faults that may be found when carrying out this test.

2. Why is it important to have ring circuit continuity?

36 Insulation resistance test

Objective: To carry out an insulation resistance test on a number of circuits.

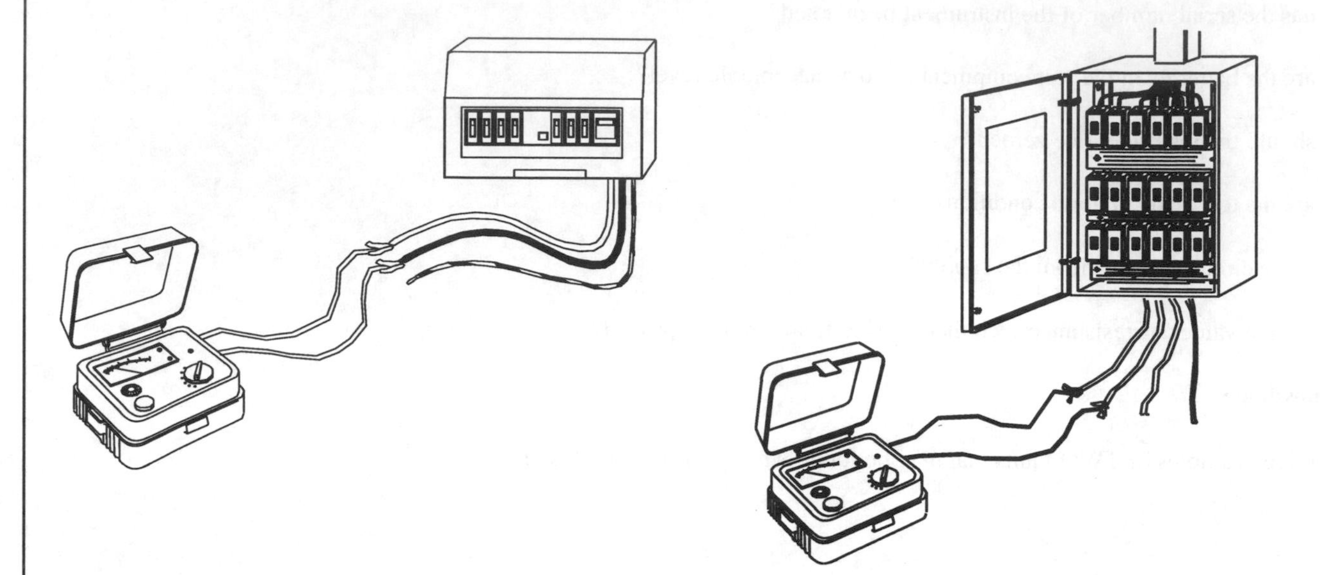

Figure 5.16

Time: related to installation being tested.

Method:

This test is to confirm that the insulation throughout the installation has not broken down.

To ensure that the insulation is capable of standing up to the stress at working pressures the test equipment uses voltages up to twice that in normal use. On installations rated up to 500 V a.c. r.m.s. it is permitted to test using 500 V d.c. and for installations rated at 500 V to 1000 V a.c. r.m.s. a test voltage at 1000 V d.c. is permitted. These voltages are accepted as being high enough to break down any poor insulation or reveal weaknesses.

Equipment:

1 insulation resistance test instrument
1 real or simulated ring final circuit

Test equipment:

Name:
Model:
Serial no.:

Results:

Fill in the relevant sections of the "Record of test results".

Points to be considered:

✓ has the correct instrument been used?

✓ has the serial number of the instrument been noted?

✓ are the batteries (if included) in the test equipment up to an acceptable level?

✓ should the instrument be zeroed?

✓ are the test leads in good condition?

✓ has the instrument been tested open circuit before use?

✓ has the instrument been tested short circuited before use?

✓ is the circuit isolated from the supply?

✓ is there any danger to any other person if the test is carried out?

✓ were all circuit switches in the ON position before testing?

✓ are the readings acceptable?

Questions:

1. List what safety precautions must be taken when carrying out an insulation resistance test at 500 V d.c.

2. The insulation resistance test instrument indicates a reading pointing to ∞ on a scale similar to that in Figure 5.17. What should be recorded on the "Record of test results"?

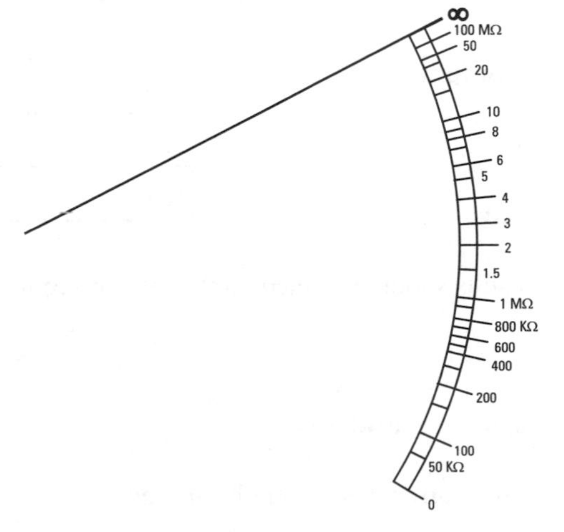

Figure 5.17

37 Polarity testing

Objective: To carry out a number of tests on an installation to verify all switches and fuses are in the phase conductor.

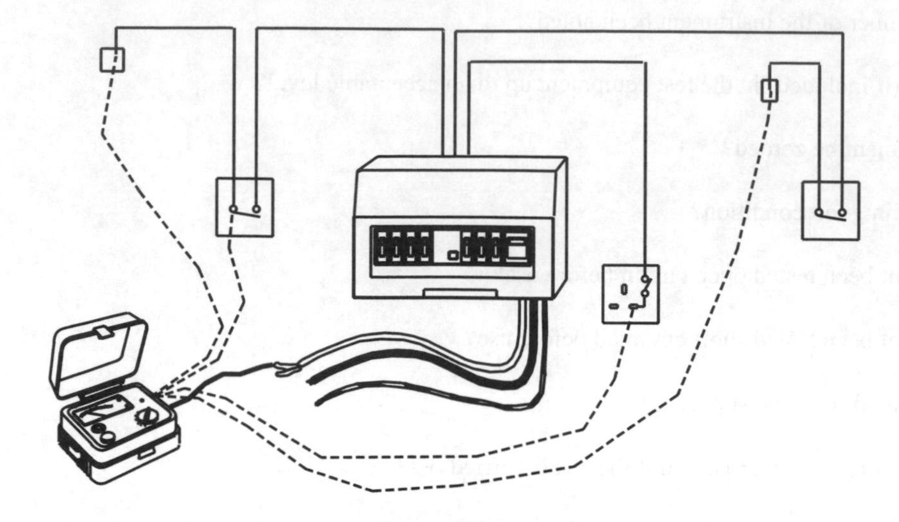

Figure 5.18

Time: related to installation being tested.

Method:

It is important that all single-pole devices are connected in the phase conductor only. These include overcurrent protection devices, switches and control contacts.

The polarity test is to confirm the correct connection of these devices.

The test can be carried out using continuity test instruments.

There is no requirement to record readings as this is a check to ensure that the installation connections have been made in the correct conductors. For future reference it should be noted that the tests were carried out and if the results were acceptable.

Equipment:

1 continuity test instrument
1 real or simulated ring final circuit

Test Equipment:

Name:
Model:
Serial no.:

Results:

The results should be noted on the "Record of test results".

Points to be considered:

✓ has the correct instrument been used?

✓ has the serial number of the instrument been noted?

✓ are the batteries (if included) in the test equipment up to an acceptable level?

✓ have the tests been carried out from a phase conductor?

✓ have all points been checked?

Questions:

1. If when the visual inspection was carried out one lamp was left in the circuit what effect, if any, could this have on your results. Give reasons for your answer.

2. On one of the circuits you are asked to test are a number of ES type lampholders. Explain how you would test these for correct polarity and what in particular you would look for?

38 Earth fault loop impedance test

Objective: To carry out an earth fault loop impedance test and check the results against the maximum values allowed for these circuits.

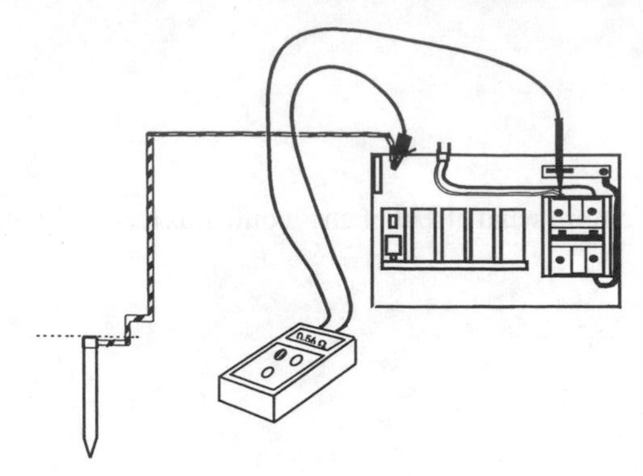

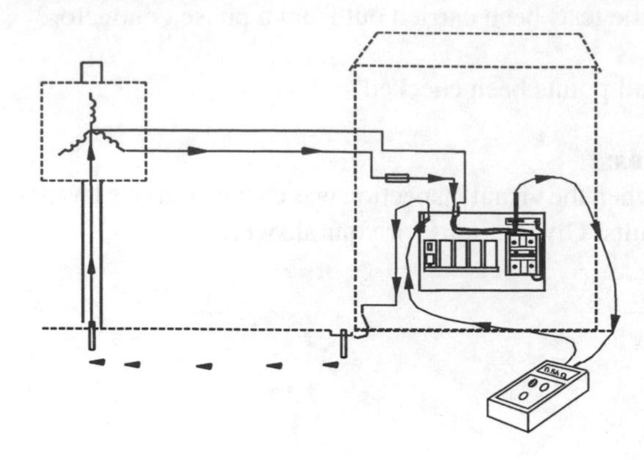

Figure 5.19

Time: related to installation being tested.

Method: This test must be carried out when the installation is connected to the Supply Company's cables.
The test can be taken from any supply point and should be carried out on all types of circuit.
There is usually some indication on the instrument that checks that there is some form of earth fault path and that the polarity is correct. If either of these checks gives a working indication do not proceed. If all is correct the test button can be pressed. This injects a current of about 25 A through the earth fault path.

Equipment:

1 earth fault loop impedance tester
1 supply suitable for carrying out the test

Test Equipment:

Name:
Model:
Serial no.:

Results:

Find out details of the circuit being tested and look up the maximum earth fault loop impedance acceptable. Fill this and the test results on the "Record of test results".

Points to be considered:

✓ has the equipment been used correctly?

✓ has the serial number of the instrument been noted?

✓ has the maximum value been looked up correctly?

✓ does the test result comply with the maximum value?

WARNING!
This exercise should be carried out in accordance with the Electricity at Work Regulations 1989 and all live conductors should be shielded and insulated from touch. All test probes should meet the requirements of the Health and Safety Executive Guidance Note GS38.

Questions:

1. Two sets of tables are given for the maximum values of earth fault loop impedance. Explain why this is so and why the values are different.

2. A circuit tested contains socket outlets and is protected by a 30 A type 2 MCB. The test impedance is 1.5 Ω. Explain what the procedure is when this happens.

39 Testing residual current devices

Objective: To carry out a test on a residual current device to determine the correct working of the device.

Equipment: 1 simulated distribution system as shown in Figure 5.20.
1 RCD test instrument

Suggested time: 30 minutes

Method:
1. Connect up the simulated distribution to a three phase supply.
2. With the supply switched ON press the test button on the RCD. If the device does not switch **OFF,** isolate the supply and check the wiring. **DO NOT PROCEED**.
3. Using the RCD test instrument plugged into the yellow phase socket outlet and switched to suitable settings, carry out the tests below and record the results.

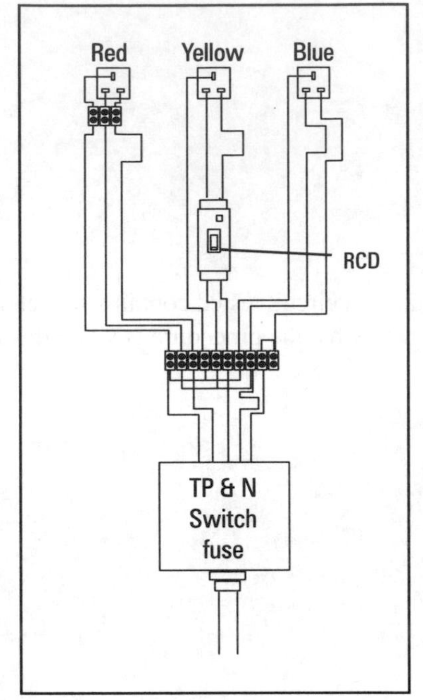

Figure 5.20

Results:

Test current	Trip time	Comments

Note: The test currents should include one below 30 mA (approximately half), one of 30 mA and one of 150 mA. The exact values may depend on the instrument used.

Question:

Considering the results does this RCD comply to BS7671?

Test equipment

Name:
Model:
Serial no.:

Remember
The earth fault loop impedance of a circuit should be kept as low as possible to allow for high fault currents.

When it is not possible to get down to the required value a Residual Current Device is required.

WARNING!
This exercise should be carried out in accordance with the Electricity at Work Regulations 1989 and all live conductors should be shielded and insulated from touch. All test probes should meet the requirements of the Health and Safety Executive Guidance Note GS38.

Questions:

1. Describe, with the aid of a circuit diagram, how a residual current device works when there is a leakage current.

2. Make up a list of situations where RCDs should be installed.

40 Voltage readings

Objective: To use different instruments to measure the same voltages and compare the results.

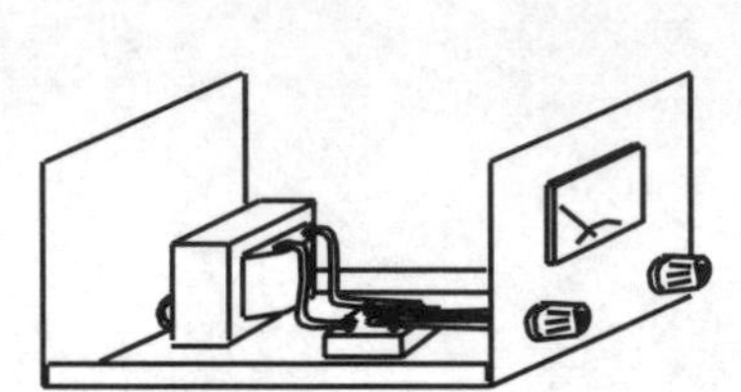

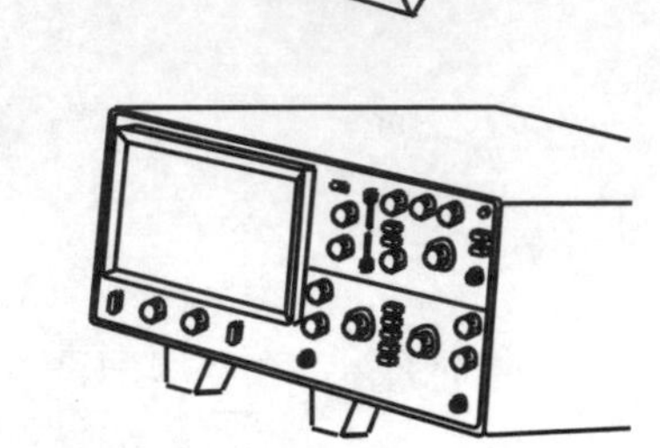

Figure 5.21

Suggested time: 1 hour

Method:

Using a constant source of supply each of the three instruments are connected across the same parts of the circuit in turn. In each case the readings are noted.

Equipment:

1 regulated d.c. P.S.U.
1 digital multimeter
1 analogue multimeter
1 cathode ray oscilloscope

Points to be considered:

✓ has safety been given top priority?

✓ has each meter been set correctly?

✓ has each meter been connected correctly?

✓ has each result been noted?

✓ have the CRO readings been calculated correctly?

Test equipment:

Digital multimeter

Name:
Model:
Serial no.:

Analogue multimeter

Name:
Model:
Serial no.:

C.R.O.

Name:
Model:
Serial no.:

WARNING!

This exercise should be carried out in accordance with the Electricity at Work Regulations 1989 and all live conductors should be shielded and insulated from touch. All test probes should meet the requirements of the Health and Safety Executive Guidance Note GS38.

Results:

Reading one – output of transformer

Meter	Voltage
Digital	
Analogue	
CRO	

Reading two – output of rectifier

Meter	Voltage
Digital	
Analogue	
CRO	

Reading three – output with battery on charge

Meter	Voltage
Digital	
Analogue	
CRO	

Questions:

1. Compare the readings and comment on the results.

2. List the advantages and disadvantages of each instrument for this type of work.

	Advantages	Disadvantages
Digital meter		
Analogue meter		
Cathode ray oscilloscope		

Record of test results

Instruments used		
	Make	Serial No.
Continuity		
Insulation		
Loop impedance		
RCD		

Installation:	Distribution board ref.	
Contractor:	External impedance (Z_E):	I_{pf}
Engineer:	Minimum insulation resistance:	
Date of tests:	Supply voltage:	

Circuit reference	Radial or ring	Protection device			Phase (colour or 3 phase)	Conductor size		Continuity			Insulation resistance MΩ		Polarity (pass or fail)	Earth loop impedance Ω		RCD			
																		Trip time (ms) at	
		Device	Type	Rating		Live	c.p.c.	L–L	N–N	c.p.c.	Phase to neutral	To earth		Design	Actual	Rated trip current (mA)	50% Y/N	100%	150 mA

6

Fault Diagnosis and Rectification

On completion of this chapter you should be able to:

- carry out fault diagnosis to locate faults in equipment and systems
- undertake necessary action to rectify faults by repair or replacement

Fault location procedure

Introduction

How do we know if there is a fault somewhere in an installation?

We can, of course, connect up the supply and see what happens. A fuse may blow, a piece of equipment start smoking, you or your colleagues may get an electric shock. Nothing may happen and it could be months or even years before something causes a fault to show up. The thing is that unless an installation is inspected and tested properly - nobody knows.

In addition to the inspection and testing that must be carried out on the completion of work, meter readings may need to be taken at other times to confirm or clarify a situation. At all times safety must be given consideration when carrying out tests. This not only applies to those personnel carrying out the test but also to anyone else who may come into contact with it.

Figure 6.1 *Removing a ceiling rose can reveal a multitude of sins!*

Before continuing read through Part 2 of Chapter 9 in "Stage 1 Design", another book in this series.

Remember
The logical approach to fault finding is:

Identify there is a fault

Obtain all the information available
(verbal, written reports, events leading up to fault)

Analyse the information/evidence
(collate information and list appropriate actions)

Sort out the options
(is there more than one?)

Select the most appropriate option
(carry out action and analyse results)

Decide - has the fault been located and identified?

If no, then analyse the results obtained, consider the information and return to the point where all the information is considered. It may then be that another option needs to be selected to solve the problem.

Remember

Follow a safe working procedure prior to undertaking fault diagnosis:

- identify load isolating devices
- prove voltage indication equipment
- secure isolation devices in "off" position
- post warning notices if appropriate
- complete relevant safety checks before restoring supply

Now let's look at a few examples before we put theory into practice and attempt to locate faults in systems and equipment.

Example

(a) Explain why it is important to carry out an inspection of an installation before any tests are carried out.

(b) List FOUR pieces of documentation required before an inspection and test can be carried out.

Answer

(a) If testing takes place without first carrying out an inspection then time and money could be wasted and dangerous situations could arise. Most of the tests use a power source of some type. If an inspection has not been carried out cables may not have been connected and could be left hanging around. When the test voltages are applied these cables could become a dangerous hazard and cause an accident.
It is also important to ensure circuits and equipment are switched correctly before any test instrument is connected. Test voltages can cause damage to equipment as well as cause accidents.

(b) Regulation 514-09-01 outlines the documentation required and includes:
  (i) Installation specification
  (ii) Plans and drawings
  (iii) Design information, showing:
    – circuit arrangements – conductors and protection devices
    – any equipment vulnerable to tests
    – isolation and switching
  (iv) Charts for recording results and comparing with design data.

Try this

The fixing screws on a 13 A socket outlet are removed so that an inspection can be carried out. List the points that should be considered.

Tip:
In Part Seven of the IEE Wiring Regulations a check list is given which should be consulted.

Using meter readings

Meter readings are often taken to confirm or clarify a situation. We have seen in the last example how this could be carried out in one set of circumstances to check design details. It is sometimes necessary to take readings to sort out a problem or confirm a suspicion. A cable starts to run warm and after tests a piece of equipment is found to be taking more current than it should. The voltage at the terminals of a distribution board is less than it was designed to be. These are examples of practical situations that need to put theory into practice to come up with a solution to the problem.

Example

A bank of fluorescent luminaires appears to be drawing more current than it was designed for. An ammeter reading indicates that a current of 8 A is flowing and the circuit voltage is 230 V at 50 Hz. A wattmeter reading shows the power is 960 W. The customer wants to know (a) what the power factor is and (b) what the current would be if the bank of fluorescent luminaires was corrected to unity power factor.

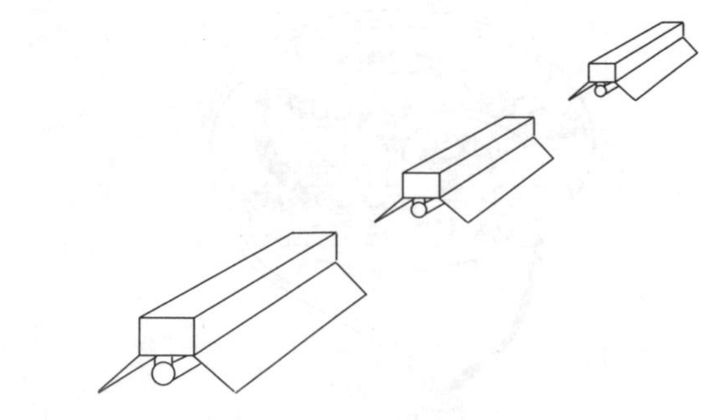

Figure 6.2 Fluorescent luminaires

Answer

(a) The power factor $= \dfrac{\text{true watts}}{\text{volt amperes}}$

$$= \frac{960}{230 \times 8}$$

$$= 0.52$$

(b) The current, if connected to unity, can be found from the wattmeter reading as at unity power factor the watts would equal the volt amperes

Current at unity $= \dfrac{\text{power}}{\text{watts}}$

$$= \frac{960}{230}$$

$$= 4.17 \text{ A}$$

Try this

A choke is tested on a bench and the following readings were obtained.

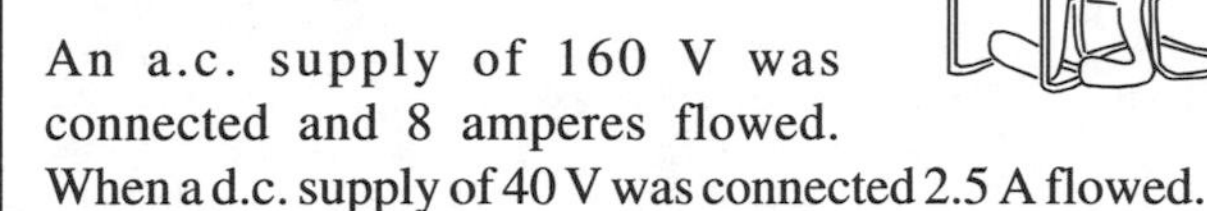

An a.c. supply of 160 V was connected and 8 amperes flowed. When a d.c. supply of 40 V was connected 2.5 A flowed.

It is required to know the power factor of the choke.

Tip:

On d.c. $\frac{U}{I}$ gives the resistance but when the same formula is used for an a.c. circuit the impedance is calculated. Once R and Z have been determined the power factor can be calculated.

Portable electrical equipment

So far we have looked at electrical installation fixed wiring and the fixed equipment associated with the wiring but the Health and Safety Executive found that about 25% of all reportable electrical accidents involved portable equipment. So it is important to consider this type of equipment when diagnosing faults. The majority of these accidents caused electric shock but many others resulted in burns from arcing or fire.

Typical accidents were caused by:

- apparatus being used for jobs for which it was not designed
- inadequate maintenance or misuse
- the use of defective apparatus

Instrument manufacturers have developed special Portable Appliance Test (PAT) equipment which is capable of carrying out all of the tests required. There are safety considerations that must be taken when using proprietary PAT instruments. As the appliances are switched on suitable precautions must be taken :

- drill bits and accessories must be removed
- appliances with moving parts should be suitably secured
- appliances with heating elements should be suitably positioned to prevent danger or heat damage

At some point in the test cycle the PAT appliance will operate and care must be taken to avoid danger, injury or damage as a result of this.

Remember

The symptoms of faults, whether inherent or developed faults, can be described as:

- a loss of supply either at the origin of the installation or localised
- an overload or fault current device operating
- transient voltages
- the failure of the insulation
- the failure of plant, equipment or component

Negligence, misuse or abuse by the installer or user may also result in faults.

Faults may also occur in an installation (other than plant and equipment) in

- cable interconnections
- cable terminators, seals and glands
- luminaires
- protective devices
- accessories such as switches or switchgear control equipment
- metering instruments
- flexible cable or cords
- portable appliances and equipment connected to a fixed installation

41 Fault location – earth continuity

Objective: To test a simulated installation for earth continuity and identify and locate any faults found.

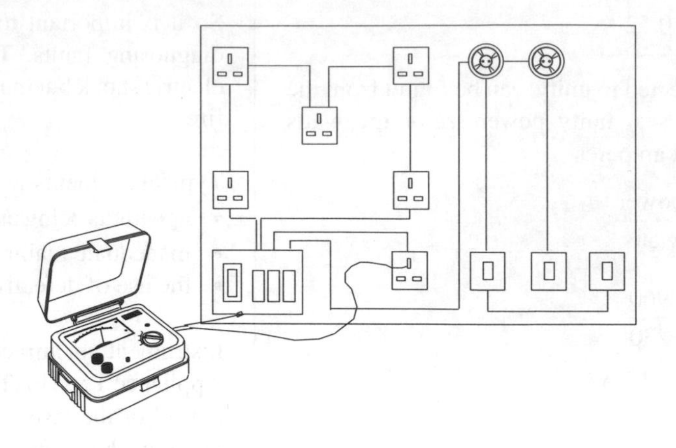

Figure 6.3

Suggested time: 30 minutes

Method: Using the appropriate test equipment carry out the recognised tests.

Equipment:

1 continuity meter
1 simulation board

Results of tests:

Findings:

Were the circuits found to be faulty?

If yes, identify the fault and give reasons for your conclusion.

Points to be considered:

✓ has any documentation/information relating to the possible fault been obtained?

✓ have any relevant people been advised that tests are to be carried out?

✓ have the correct procedures for isolation been carried out?

✓ have the tests been carried out in a safe manner?

✓ has the fault/s been diagnosed correctly?

✓ has the correction procedure been agreed with relevant people, using the appropriate tools, equipment and materials?

✓ has the installation been left in a safe condition?

✓ if the installation has been modified have relevant people been notified?

42 Fault location on a ring final circuit

Objective: To test a ring final circuit and identify and locate any faults found.

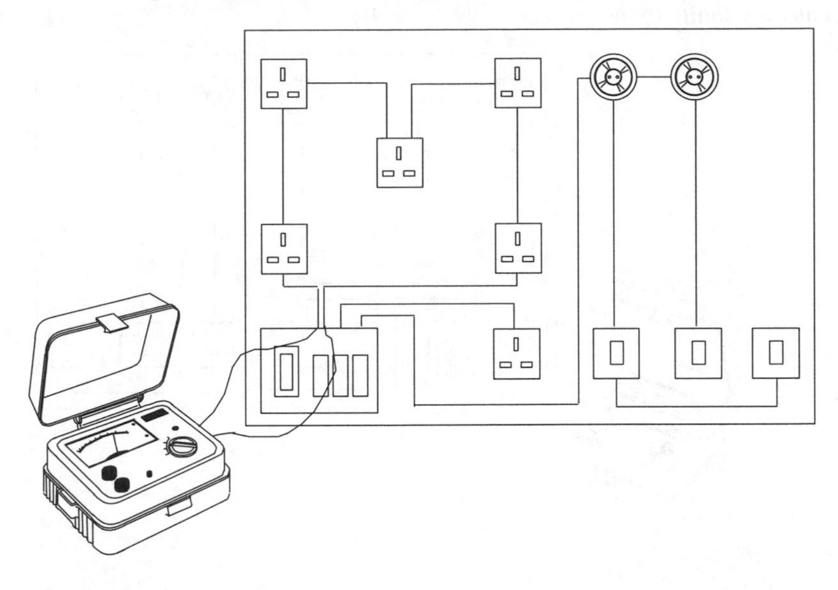

Figure 6.4

Suggested time: 30 minutes

Method: Using the appropriate test equipment carry out the recognised tests.

Equipment:

1 milliohmmeter
1 simulation board

Results of tests:

Findings:

Were the circuits found to be faulty?

If yes, identify the fault and give reasons for your conclusion.

Points to be considered:

✓ has any documentation/information relating to the possible fault been obtained?

✓ have any relevant people been advised that tests are to be carried out?

✓ have the correct procedures for isolation been carried out?

✓ have the tests been carried out in a safe manner?

✓ has the fault/s been diagnosed correctly?

✓ has the correction procedure been agreed with relevant people, using the appropriate tools, equipment and materials?

✓ has the installation been left in a safe condition?

✓ if the installation has been modified have relevant people been notified?

43 Fault location – insulation resistance

Objective: To test a simulated installation for insulation resistance and identify and locate any faults found.

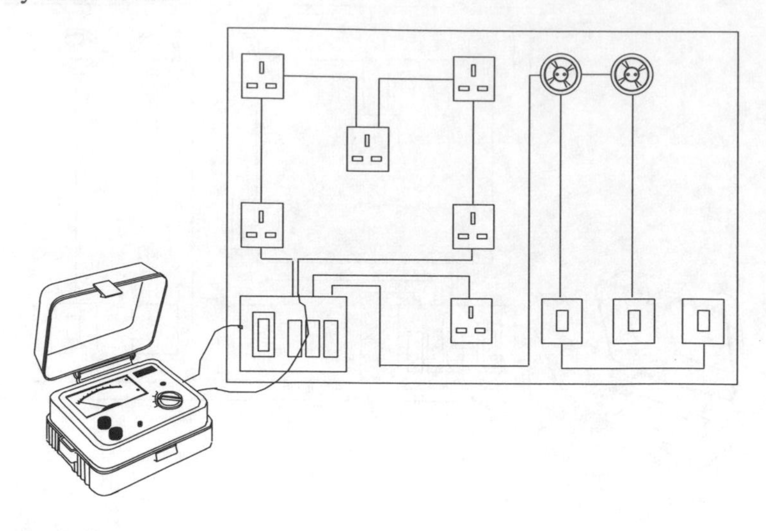

Figure 6.5

Suggested time: 30 minutes

Method: Using the appropriate test equipment carry out the recognised tests.

Equipment:

1 insulation resistance meter
1 simulation board

Results of tests:

Findings:

Were the circuits found to be faulty?

If yes, identify the fault and give reasons for your conclusion.

Points to be considered:

✓ has any documentation/information relating to the possible fault been obtained?

✓ have any relevant people been advised that tests are to be carried out?

✓ have the correct procedures for isolation been carried out?

✓ have the tests been carried out in a safe manner?

✓ has the fault/s been diagnosed correctly?

✓ has the correction procedure been agreed with relevant people, using the appropriate tools, equipment and materials?

✓ has the installation been left in a safe condition?

✓ if the installation has been modified have relevant people been notified?

44 Fault location – polarity

Objective: To test a simulated installation for polarity and identify and locate any faults found.

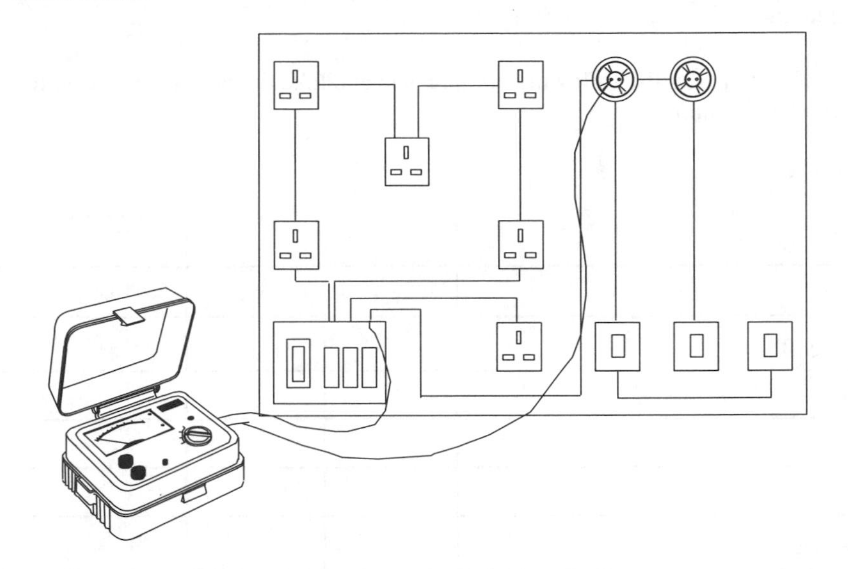

Figure 6.6

Suggested time: 30 minutes

Method: Using the appropriate test equipment carry out the recognised tests.

Equipment:

1 continuity meter
1 simulation board

Results of tests:

Findings:

Were the circuits found to be faulty?

If yes, identify the fault and give reasons for your conclusion.

Points to be considered:

✓ has any documentation/information relating to the possible fault been obtained?

✓ have any relevant people been advised that tests are to be carried out?

✓ have the correct procedures for isolation been carried out?

✓ have the tests been carried out in a safe manner?

✓ has the fault/s been diagnosed correctly?

✓ has the correction procedure been agreed with relevant people, using the appropriate tools, equipment and materials?

✓ has the installation been left in a safe condition?

✓ if the installation has been modified have relevant people been notified?

45 Faulty direct-on-line starter

Objective: To undertake necessary action to rectify faults by repair or replacement.

Suggested time: 2 hours

Method: Test and inspect the following components and take the necessary action to rectify the faults you have found.

Tick the appropriate boxes:

		Yes	No	Action taken	
				Repaired	Replaced
A	**Contactor coil**				
(i)	open circuit				
(ii)	earth fault				
(iii)	short circuit				
B	**Main contacts**				
(i)	phase to earth fault				
(ii)	open circuit				
(iii)	short between phases				
C	**Remote stop/start**				
(i)	short circuit start button				
(ii)	earth fault				
(iii)	open circuit stop button				
D	**Overload assembly**				
(i)	open circuit one coil				
(ii)	faulty auxiliary contact				
(iii)	incorrect overload setting				
E	**Hold-on (retaining) contact**				
(i)	high resistance (dirty) contact				
(ii)	open circuit				
(iii)	short circuit				

Equipment: Three similar 3-phase D-O-L starters (each wired with remote stop/start controls).
Spare components as required.
One voltage indicator conforming to GS38.
One continuity tester with ohms range.
One insulation resistance tester.
One isolation notice.

Note: The starters have different faults. The tutor will select which starter the candidate will work on.

Points to be considered:

✓ have safe isolation procedures been carried out correctly?

✓ have all safety precautions been taken?

✓ have the faults been identified correctly?

✓ has the correct action been taken?

✓ if appropriate, have the repairs and their costs been agreed with relevant people?

✓ has the test equipment been proven?

7

Revision Exercises

This chapter has been included to help with revision for courses such as the City and Guilds Electrical Installation Course 2360 Part 2, the City and Guilds Electrical Installation Technology Course 8230 Part 2 (overseas) and appropriate SCOTVEC and BTEC courses. Section one gives guidance on why, where and how to revise and includes example questions and answers followed by a similar question for you to try. Section two contains more short answer questions and a multi-choice question paper for you to check on the areas where you still need to revise.

Section one

Why do you revise?

The decision as to whether or not to revise is in the end up to each individual. Some people will argue that there should be no need to revise, others think it will make up for a lack of study at earlier stages. Although everybody is different and has their own requirements, their ultimate aim is usually the same and that is to pass the examination.

It is often quite a long time from the start of a course of study to the time of the examination. In that time new subjects have been introduced and details or original points forgotten. Some areas of study may also have been misunderstood. Revision should be seen as a way of bringing subjects back to mind and putting them in some order.

Revision should not be seen as a substitute for original study but should be used to go over work covered earlier and to fill in any gaps.

It is always better to turn up on the day of the examination feeling confident, and good, well-planned revision can help.

When do you revise?

In most cases the night before the examination is too late to start revising. Similarly, before the subject has been covered is too early. Almost anytime in between may be suitable.

As everybody is different, it is not possible to say what day of the week or what time of day is best to revise. There are, however, some points that should be considered. It takes time to revise properly and that time has to be found and set aside. Revision also requires concentration and this usually means finding a quiet place where there will be little or no interference. Ideally, the revision should be carried out when the mind is fresh and responsive, not late at night when it is fighting to stay awake.

How do you revise?

Before starting to revise it is important to draw up a programme of work. To help to determine which subject areas are required the questions at the front of each section should be attempted without referring to the rest of the section. This exercise will also give an indication as to the depth that each subject area needs revising to.

A good revision programme should include the subject area to be covered, the type of questions that should be answered and a target time each part should take to complete.

The following guide may help when planning a revision session using this series of books:

- Select the subject area to revise
  - > go to that chapter in the book
- Select an appropriate question from the ones later on in this chapter
  - > try the question on rough paper
  - > go to that area in the relevant book
- Read through the background theory
  - > check your answer for inaccuracies and/or omissions
  - > look up information in other books if necessary
  - > make notes to remember facts
- Attempt a similar question
  - > try to answer the question on rough paper
  - > make a note of the "key points" so that you can refer to them when required
  - > look up information in other books if necessary
  - > complete the answer
- Select another similar question from the end of the section
  - > look up details
  - > answer the question

As a check on which topics need further revision, try the 50 multi-choice questions on p.111.

The examination

There are a number of other points that should be considered when revising that can help when taking the examination. Where books such as the "IEE Wiring Regulations" and "On Site Guide" can be used in the examination room these need to be studied when revising so that the relevant information can be found quickly. Even with these publications available there are going to be some facts that have to be remembered. It can be helpful to build a story up around them, fit them into a rhyme or visualise them in picture form. This applies equally to formulae and mathematical equations.

One final point - good clear presentation with labelled diagrams could make just the difference between a pass and a failure!

A number of examples will now follow with answers similar to those required in an examination. Then there will be a similar question for you to attempt. The answers given are for guidance only and are not necessarily the only possible solutions.

Example 1

(a) A delta/star transformer supplies a factory at 400 V three-phase and 230 V single phase. Draw a labelled diagram to show how the following can be obtained from the star connected transformer winding.

i) 400 V, three-phase 3 wire
ii) 400 V, three-phase 4 wire
iii) 230 V, single phase
iv) 400 V, single-phase

(b) Calculate the line current in each phase if a balanced load of 75 kW at a power factor of 0.75 lagging is supplied with 400 V three-phase.

(Reference Stage 1 Design, Chapter 1, Intermediate Science and Theory Chapter 5)

Answer

(a) The supplies obtained from a delta star transformer.

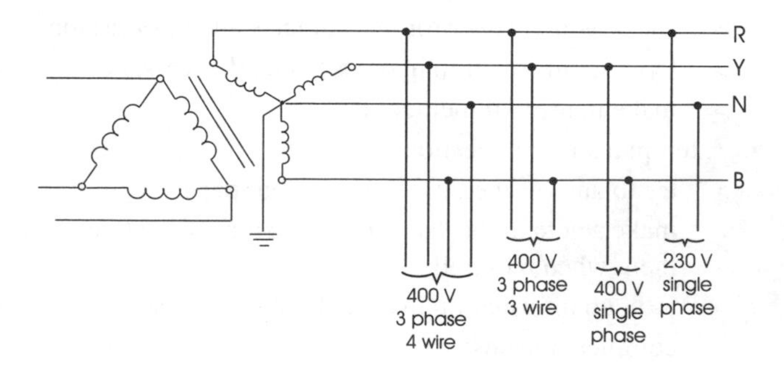

Figure 7.1

(b) Power in three-phase:

$$P = U I \sqrt{3} \cos \phi$$

$$I = \frac{P}{U \sqrt{3} \cos \phi}$$

$$I = \frac{75 \times 1000}{400 \times 1.73 \times 0.75}$$

$$= 144.51 \text{ A}$$

Try this

(a) Four single-phase loads are to be connected to a three-phase 400–230 V star connected transformer. Three of these are to supply 230 V loads but the fourth is to supply a 400 V load. Draw a diagram to show how the voltages for each load can be obtained.

(b) Calculate the power, in kilowatts, of one 230 V single phase load if a current of 22 A at a power factor of 0.85 is flowing.

Tips:

(a) Notice that all four loads are single-phase so each will have two wires supplying it.

(b) Remember that $P = U I \cos \phi$ where $\cos \phi$ is the power factor.

Example 2

The 230 V supply to a small market garden is by 2 single phase overhead cables.

Explain:

(a) how the consumer's installation should be protected from possible earth faults

(b) the difference between the earthing arrangements of this and that of a TN-C-S system.

(Reference Stage 1 Design, Chapter 2)

Answer

(a) As the supply to the market garden is by two cables only, the supply company does not include a connection to earth. The supply is in effect a TT system and the consumer has to provide the earth fault protection. This will consist of an RCD and an earth electrode in addition to the overcurrent protection equipment.

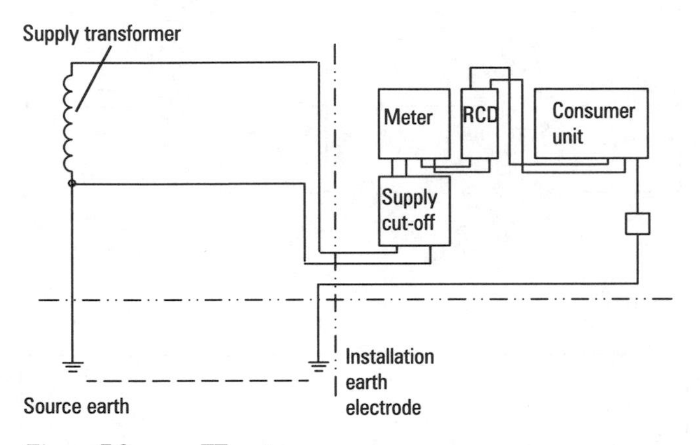

Figure 7.2 *TT system*

Under fault conditions the current flows through the phase conductor but returns through the protective conductor not the neutral. The RCD detects the phase and neutral currents are not the same and automatically switches off the electrical supply.

(b) As we have seen in (a) the supply company does not include a connection to earth with a TT system. A TN-C-S system is supplied with a cable where the earth connection and the neutral conductor are combined. At the consumer's premises these are separated so that as far as the consumer is concerned it is a phase, neutral and earth supply. It is not essential to include an RCD with all TN-C-S systems but they are sometimes added for extra protection.

Try this

Draw the sequence of control equipment in a consumer's premises for a TN-S system. Include and label the supply company's and consumer's equipment.

Tip:
A diagram showing a TN-S system can be found in Part 2 of BS 7671 but don't just copy this – draw it in your own way.

Example 3

(a) Draw the circuit diagram for a switch start fluorescent luminaire.

(b) Explain the function of each component shown in the circuit for (a).

(Reference Intermediate Science and Theory, Chapter 8)

Answer

(a)

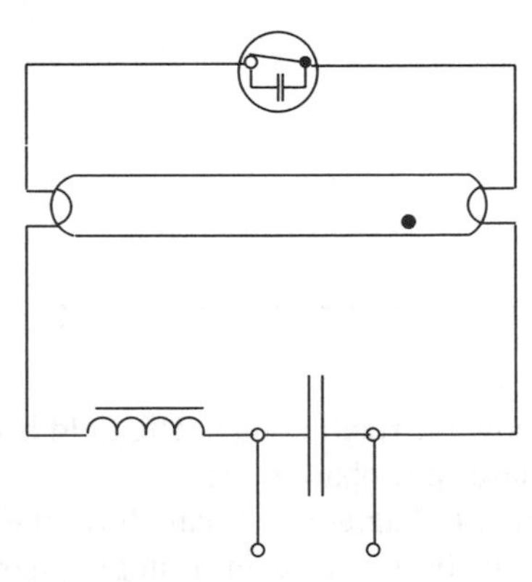

Figure 7.3 *Circuit diagram for a switch start fluorescent luminaire*

(b) When the supply is connected to the circuit, current flows through the choke, through the lamp filament A, the starter, the second lamp filament B and back to the supply. The gas in the starter heats up and the bimetal contacts close. The gas now cools down and the contacts open. During this time the two filaments in the tube have heated up and an electron cloud has been created around them. The opening of the contacts in the starter stops the current flow causing a collapse of magnetic field in the choke, inducing a high voltage. This voltage strikes the tube and starts the electron flow through the low pressure mercury vapour. The low resistance of the tube now shorts out the starter switch. Once the tube has a discharge flowing through it, the choke becomes a current limiting device.

If the lamp is switched off and then back on again the procedure is repeated and the lamp will strike almost immediately.

The power factor correction capacitor brings the current back into line with the voltage so that the current in the circuit conductor is kept to a minimum.

Try this

If the power factor correction capacitor is removed from a discharge lighting circuit what will the effect be on:

(a) the starting of the lamp
(b) the current flowing in the luminaire
(c) the current flowing in the circuit supplying the luminaire

Tip:

Remember that the power factor correction capacitor is there to counteract the inductive effect of the choke.

Example 4

(a) Explain how the rotating magnetic field is produced in a single-phase split-phase motor.
(b) Using circuit diagrams explain how the direction of rotation can be reversed in a single-phase split-phase motor.

(Reference Intermediate Science and Theory, Chapter 6)

Answer

(a) The single-phase supply by itself will not produce a rotating magnetic field to get the motor started. A second winding has to be put in the stator core, connected in parallel with the first.

The main winding is made of thick wire and is placed deep in the iron core. The second coil is made of thinner wire and placed nearer to the surface of the iron core. As the main winding has less resistance and more reactance than the other coil a phase shift is produced. This in effect produces two phases inside the motor.

When the motor is switched on the phase difference in the two windings is enough to produce a rotating magnetic field to get the motor started. As the motor reaches full speed a centrifugal switch disconnects the second (start) winding as this is no longer required.

(b) To reverse the direction of rotation of a single-phase motor the connections to either the start or run winding can be changed over, but not both.

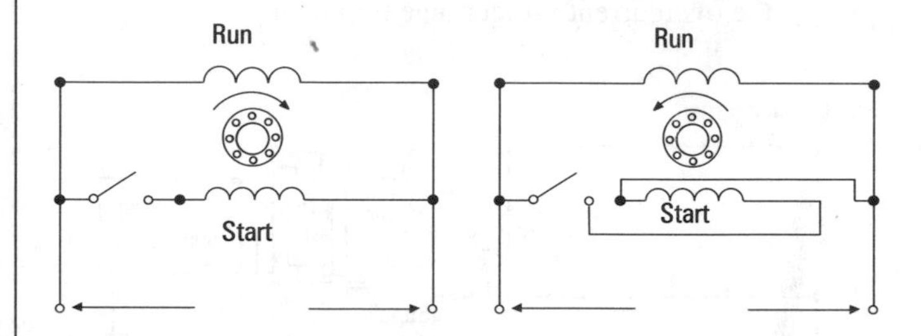

Figure 7.4

Try this

Explain using diagrams how the direction of rotation can be reversed on a three-phase cage rotor induction motor.

Tip:

Remember the sequence from the supply is always Red, Yellow, Blue. The motor will follow this sequence.

Example 5

Three resistive single-phase loads are connected to a three-phase distribution board. The loads are:

Red phase 12 A

Blue phase 5 A

Yellow phase 7 A

Using a scale of 10 mm = 1 A construct a phasor diagram and measure the current flowing in the neutral of the supply to the distribution board.

(Reference Intermediate Science and Theory, Chapter 4)

Answer

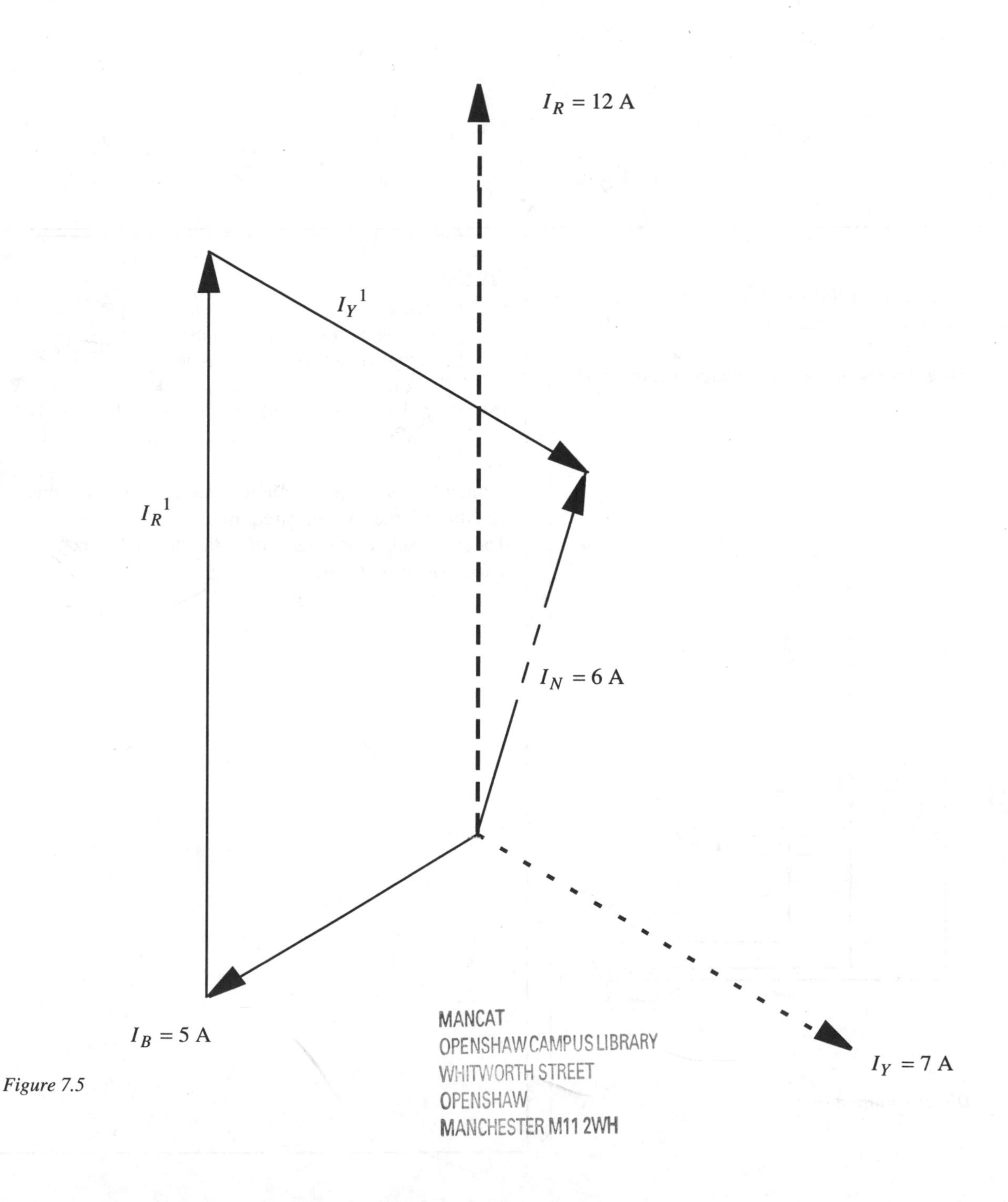

Figure 7.5

Try this

A three-phase star connected transformer is loaded so that there is 60 A drawn from the red phase with 40 A from the yellow and 100 A from the blue. Using a scaled phasor diagram determine the current that would be flowing in the neutral return to the start point of the transformer.
Assume the power factor in all loads to be unity.

Example 6

Draw the circuit diagram for a 400 V three-phase motor started from a direct-on-line starter with a 230 V coil. Label all parts.

(Reference Intermediate Science and Theory, Chapter 6)

Answer

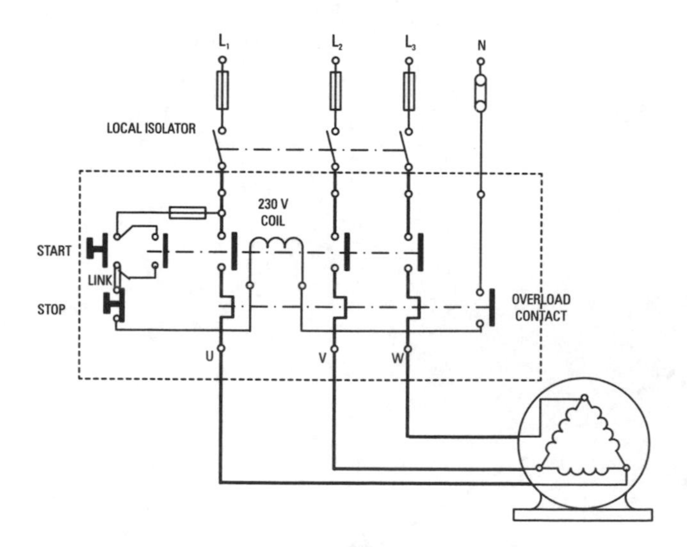

Figure 7.6 Direct-on-line starter

Try this

(a) Explain why overcurrent protection in motor starters must be designed so that a time delay is built in.

(b) Describe, with the aid of a labelled diagram, one example of how this can be achieved.

Tip:

Remember that when a motor starts from a stationary position a large current is required.
The overload devices use either the thermal or magnetic effects of current flow.

Revision Exercises

Section two

Supply systems

1. A 5 kW load takes a current of 40 A when connected to a 230 V supply. Calculate the power factor of the load.
2. A 400 V three-phase supply is connected to a delta connected heating load. The rating of each of the three elements is 8 kW. Calculate the current flowing in each supply cable.
3. Explain, with the aid of a diagram, the earthing arrangements of a TN-C-S system.
4. What is the alternative name for an isolator?
5. Explain the advantage of a rising main busbar system over a conduit system to supply a number of suites of offices.
6. Draw a scaled phasor diagram and determine the current flowing in the neutral conductor when three single-phase heating loads are connected to a 400 V 4 wire supply. The three loads are:

Red phase	4 kW
Yellow phase	3 kW
Blue phase	5 kW

7. Draw the circuit diagram so that a 9 V full wave rectified supply can be obtained from a 230 V 50 Hz supply using:
   1 double wound transformer with a centre tap on the secondary winding

   2 single diodes

Wiring systems

1. Select a suitable wiring system for the following areas associated with petrol filling stations. Give reasons for your choice.
   (a) sales shop which is outside any hazardous area
   (b) canopy lighting 3 m above the top of the pumps
   (c) pump supplies
2. A 230 V, 12 kW domestic electric cooker is supplied by a 15 metre run of PVC insulated and sheathed cable clipped to a surface. The cooker control unit incorporates a 13 A socket outlet. If the circuit is protected by a BS 3036 fuse determine the minimum size of cable which may be used.
3. A radial circuit supplying a 13 A socket outlet is wired in 2.5 mm^2 cable which has a 1.5 mm^2 circuit protective conductor incorporated within the sheath. Calculate:
   (a) the R1 and R2 value if the cable is 18 m long
   (b) the actual Zs value if Ze is 0.3 Ω
   (c) the maximum Zs value if a 20 A protection device to BS EN 60269-1 is used
4. (a) What is the minimum fault current that needs to flow through a 100 A fuse to BS EN 60269-1 if it is to operate in 0.4 seconds?
   (b) If the protection device in (a) was replaced with a fuse to BS 3036 when the same fault current was flowing how long would it take the new fuse to operate?
5. Draw a block diagram showing how the control for a space heating system would work. the controller is supplied with information from a room thermostat, cylinder thermostat and programmer. It has to control the fuel supply to the boiler and the pump to the hot water.

A.C. motors

1. Explain, with the aid of diagrams, how the following motors can have their direction of rotation reversed:
   (a) universal (series motor)
   (b) three-phase wound rotor motor
2. Describe how the windings of a totally enclosed motor are cooled.
3. Some motors have thermistors embedded in their stator windings. Explain how these can prevent a motor from being damaged due to overheating.
4. When working on full load a 230 V, 50 Hz single-phase induction motor has a power of 1.75 kW. If the power factor of the motor is 0.7 what is the input current?
5. The resistance of the windings of a 230 V single-phase motor is 9 Ω with an inductive reactance of 12 Ω. A capacitor with an Xc of 24 Ω is connected across the motor for power factor correction. Calculate:
   (a) the impedance of the motor winding
   (b) the current flowing through the motor winding
   (c) the current flowing through the capacitor
   (d) the total current taken from the supply
6. A motor has a full load power of 10 kW when connected to a 400V three-phase supply. If the motor efficiency is 75% and power factor is 0.8 calculate the current flowing in its supply cables.
7. A washing machine induction motor has two sets of windings within it so that two speeds can be obtained. For the washing action the motor is connected as a 4 pole machine and has a rotor speed of 1450 rev/min. When it is being used for spin drying the motor is used as a 2 pole machine with a rotor speed of 2800 rev/min. If it is connected to a 230 V 50 Hz supply what are the percentage slips at each speed?

D.C. Machines

1. State:
   (a) how the field current of a d.c. shunt wound motor would normally be varied
   (b) ONE effect of reducing this field current
2. Describe briefly why a "faceplate" starter is connected into a d.c. shunt motor circuit.
3. Draw a circuit diagram for a "short-shunt" compound d.c. motor.
4. State THREE factors which would affect the e.m.f. induced in the armature of a d.c. shunt generator.
5. (a) Redraw the diagram below and sketch the resulting magnetic flux paths on the diagram
   (b) Give TWO practical applications for this arrangement:

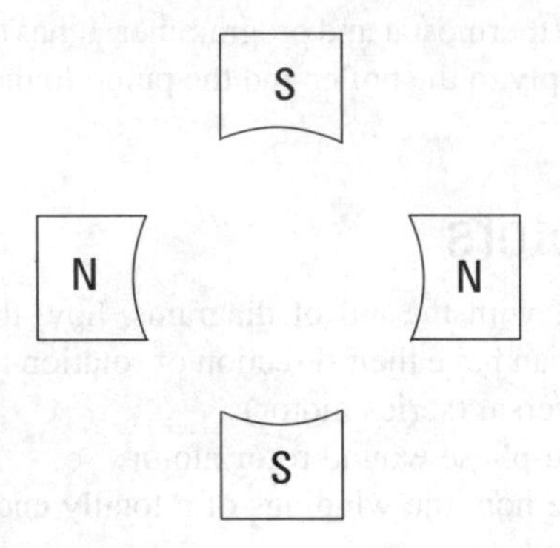

Transformers

1. (a) What are the TWO different types of losses that will occur in a transformer on load?
   (b) Which of these losses does not change when the load changes?
2. A double-wound single-phase transformer has a primary to secondary turns ration of 1:10. Calculate the:
   (a) primary voltage needed to supply a load at 200 V
   (b) primary current if the load current is 10 A
3. Draw a circuit diagram of a 230/110 V on-site transformer suitable for supplying portable tools
4. A single-phase 230/12 V transformer has a 12 V 60 W lamp connected to its secondary circuit. Determine the current in the primary circuit.
5. A double-wound transformer supplied with 20 A at 230 V supplies a load of 50 A at a terminal voltage of 100 V. Calculate the efficiency of the transformer.

Lighting

1. (a) Draw a labelled circuit diagram for a 150 W low pressure sodium lamp.
   (b) Describe the starting process for this type of lamp.
2. (a) Explain why the lamp ratings of fluorescent tubes cannot be used alone when calculating the load current of a circuit.
   (b) What is the minimum number of circuits that can be used with the loads shown below if 5 A protection devices are used?
   15 twin lamp 60W fluorescent luminaires
   24 single lamp 70W fluorescent luminaires
3. (a) Explain what is meant by stroboscopic effect when related to discharge lighting.
   (b) Describe what will happen if the "glow" type starter is removed from a fluorescent luminaire when the tube is lit.
4. An 800 cd lamp is suspended 4 m directly above a workbench. Calculate the illumination
   (a) at a point 4 m directly below the lamp
   (b) at a point 3 m away from the point in (a) on the same workbench
5. Calculate the luminous flux required to provide an illumination level of 200 lux in a room 6 m × 8 m if the C of U and L.L.F. are 0.6 and 0.8 respectively.

Inspection and testing

1. An inspection reveals that a protection device to BS 88 has been replaced with one to BS 3036. Explain what effect this could have on the circuit in the event of:
   (a) an overload
   (b) a short circuit
2. Explain the procedure for carrying out an insulation resistance test from a consumer unit before the supply is connected.
3. Describe how an earth fault loop impedance test can be carried out on a single-phase motor circuit.
4. When could a Minor Electrical Installation Works Certificate be used?
5. State the test sequence for carrying out an insulation resistance test on a three-phase installation.

Multi-choice questions

Write the letter of the correct answer in the box provided in the answer grid at the end of this paper.

1. An employer is required to report an accident to the Health and Safety Executive if an employee is
   a. treated at work for a cut hand
   b. sent home for 24 hours due to the accident
   c. unable to work for more than three days due to the accident
   d. at a hospital casualty department for a morning before returning to work
2. If the electricity supply company states that the nominal supply voltage is 230 V, the minimum voltage they can legally supply under the current arrangement is
   a. 220 V
   b. 216.2 V
   c. 224 V
   d. 225.6 V
3. The supply voltage to a substation is 11 kV and has a line current of 100 A to delta connected windings. The current flowing in the phase winding of the transformer is
   a. 50 A
   b. 57.8 A
   c. 100 A
   d. 173.2 A
4. A high breaking capacity fuse has a British Standard number of
   a. BS EN 60898
   b. BS3036
   c. BS1361
   d. BS EN 60269-1
5. A table lamp has a 5 A BS 1362 fuse in the plug which is in a 13 A socket protected by a 30 A Type 2 mcb. If the consumer unit is protected by a 100 A fuse, which is the fuse that should operate when the flex on the lamp shorts out?
   a. 5 A
   b. 13 A
   c. 30 A
   d. 100 A
6. A domestic cooker is rated at 12 kW when connected to a 230 V supply. If diversity is allowed at 10 A + 30% full load in excess of the 10 A, the assumed current demand would be
   a. 32.22 A
   b. 22.6 A
   c. 45.22 A
   d. 52.17 A
7. A single-phase motor is connected so that its voltage, current and wattage can be monitored. One set of readings gives V = 230 V, I = 6.25 A and P = 1.15 kW. The power factor in this case is
   a. 0.5
   b. 0.6
   c. 0.8
   d. 1.25
8. If a supply is said to be a TN-C-S system the protective and neutral conductors are
   a. combined throughout the supply and consumer's premises
   b. separate throughout the supply and consumer's premises
   c. combined to the consumer's premises then separate
   d. separate to the consumer's premises and then combined
9. A triple pole and neutral isolator consists of
   a. one triple pole and one single pole switch
   b. one triple pole switch and a neutral link
   c. two double pole switches
   d. one four pole switch
10. Which of the following is correct for the bridge rectifier?

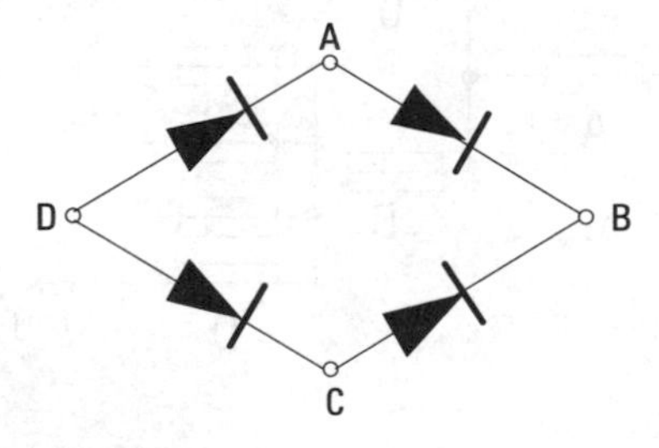

	A	B	C	D
a.	d.c.–	a.c.	d.c.+	a.c.
b.	d.c.–	d.c.+	a.c.	a.c.
c.	a.c.	a.c.	d.c.+	d.c.–
d.	a.c.	d.c.+	a.c.	d.c.–

11. If each of the cells shown in the diagram has an internal resistance of 0.03 Ω the total resistance of the battery would be

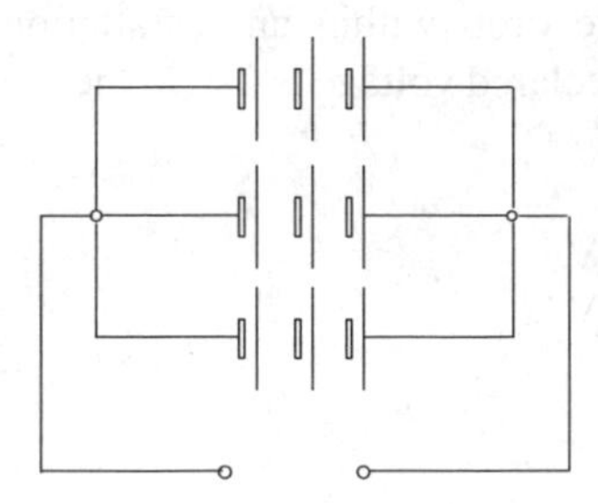

   a. 0.01 Ω
   b. 0.03 Ω
   c. 0.09 Ω
   d. 0.27 Ω

12. The fuse recommended for use in the test probes of a voltmeter should not exceed
    a. 30 mA
    b. 500 mA
    c. 1 A
    d. 2.5 A
13. A meter is designed to read full scale when 100 V is put across it. The turns ratio of a voltage transformer, required so that this meter can be used to monitor the voltage on supply with a maximum of 11 kV, will be
    a. 9:1
    b. 10:1
    c. 90:1
    d. 110:1
14. A suitable wiring system for a cow shed would be
    a. galvanised steel conduit
    b. black enamel steel conduit
    c. high impact PVC conduit
    d. bare MIMS cable
15. The correction factor of 0.725 is applied when a protection device is installed to
    a. BS EN 60269-1
    b. BS1361
    c. BS EN 60898
    d. BS3036
16. If a fault develops as indicated the protection device that should operate is

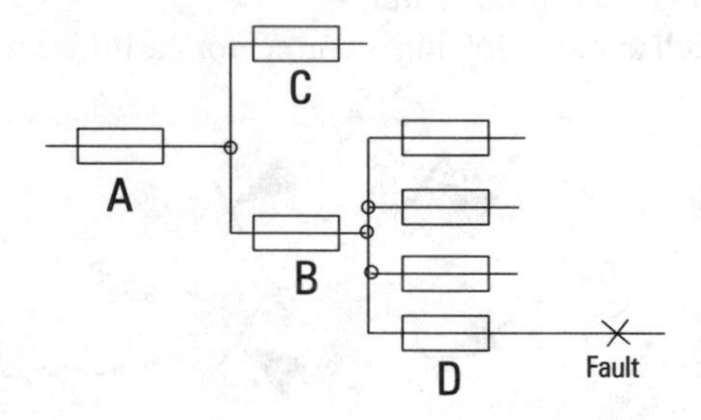

    a. A
    b. B
    c. C
    d. D
17. The correction factor identified as Ct is for
    a. ambient temperature
    b. operating temperature of the conductor
    c. thermal insulation
    d. two groups of conductors
18. In conditions where the supply voltage falls by 6% and the voltage drop within an installation is 4%, if the nominal declared voltage is 230 V the voltage at the load is
    a. 230 V
    b. 220.8 V
    c. 216.2 V
    d. 207 V
19. A cable has a tabulated voltage drop of 18 mV/A/m. If the cable is 25 m long, carries a total load current of 17 A and is protected by a 20 A BS EN 60269-1 fuse the maximum voltage drop at the load is
    a. 7.65 V
    b. 8.50 V
    c. 9.00 V
    d. 15.30 V
20. A circuit has a design current of 28A and is protected by a 32 A BS EN 60269-1 fuse. If correction factors for ambient temperature 0.82 and grouping 0.65 then the minimum current-carrying capacity of the cable supplying the circuit would be
    a. 28 A
    b. 32 A
    c. 50 A
    d. 60 A
21. The minimum permissible cross sectional area for aluminium conductors is
    a. 2.5 mm^2
    b. 4.0 mm^2
    c. 6.0 mm^2
    d. 16.0 mm^2
22. The total earth fault loop impedance path can be calculated from
    a. Ze = Zs + R1 + R2
    b. Zs = Ze + R1 + R2
    c. Ze = Zs + R1 – R2
    d. Zs = Ze + R1 – R2
23. A BS 3036 fuse rated at 20 A when carrying a fault current of 100 A will operate in
    a. 1.8 seconds
    b. 8.5 seconds
    c. 0.8 seconds
    d. 1.5 seconds
24. The current carrying conductor in the diagram will move

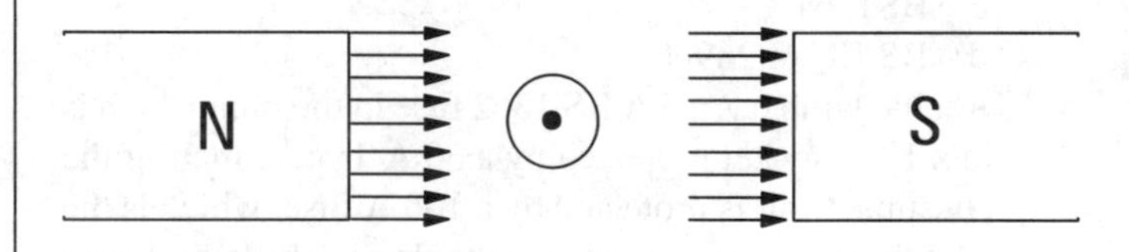

    a. downwards
    b. upwards
    c. to the left
    d. to the right
25. A three-phase induction motor can have the direction of rotation reversed by reversing
    a. the start windings
    b. the run windings
    c. any two phases
    d. all three phases
26. The synchronous speed of a six pole induction motor operating from a 50 Hz supply is
    a. 500 rev/min
    b. 1500 rev/min
    c. 600 rev/min
    d. 1000 rev/min

27. The total current I_T of the circuit shown in the diagram is

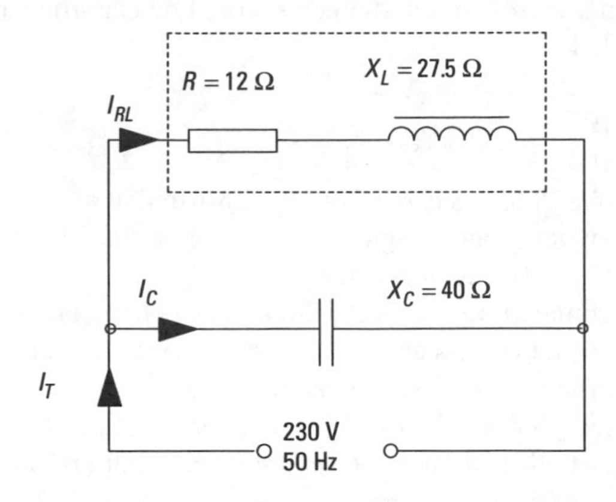

a. 3 A
b. 10 A
c. 14 A
d. 16 A

28. A 400 V, three-phase motor has an input power of 5 kW and a power factor of 0.75 lagging. The line current taken by the motor is
a. 11.75 A
b. 9.6 A
c. 16.67 A
d. 12.5 A

29. The purpose of the start winding of a single-phase motor is to
a. increase the circuit resistance
b. reduce the starting power
c. produce a starting torque
d. reduce the power factor

30. The appropriate type of starter for starting a three-phase 2.75 kW cage rotor motor against a very light load is
a. auto-transformer
b. rotor resistance
c. direct-on-line
d. star delta

31. A motor is stamped with the symbol shown. An application for such a motor could be

a. cooling electronic equipment
b. central heating pump
c. in a freshwater well
d. petrol pump

32. The power in a three-phase delta connected circuit may be calculated from:
a. $P = \sqrt{3}\ UI \cos \phi$
b. $P = \dfrac{UI}{\sqrt{3} \cos \phi}$
c. $P = UI \cos \phi$
d $P = \dfrac{\sqrt{3}}{UI \cos \phi}$

33. The starting mechanism of a capacitor start split-phase motor consists of start windings in series with the
a. run windings, centrifugal switch and capacitor
b. centrifugal switch and run windings
c. centrifugal switch and capacitor
d. run windings and capacitor

34. Lumens per watt is the measurement of
a. luminance intensity
b. luminous efficacy
c. luminous flux
d. luminance

35. If X = 24, R = 18 the value of Z must be

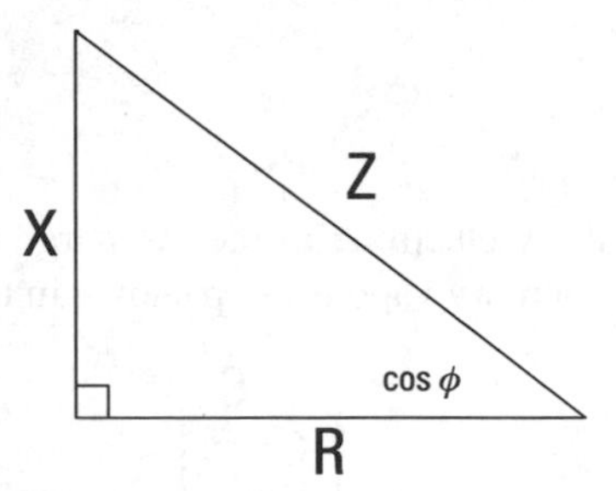

a. 24
b. 30
c. 36
d. 42

36. A capacitor can be connected across the supply to a discharge lamp to
a. reduce the current flowing through the lamp control equipment
b. improve the power factor of the lamp and control gear
c. reduce radio interference due to the lamp starting
d. smooth out the supply to the lamp

37. A GLS lamp has an efficacy between
a. 10 to 18 lm/watt
b. 32 to 58 lm/watt
c. 55 to 120 lm/watt
d. 60 to 78 lm/watt

38. Inductive reactance is measured in
a. inductors
b. farads
c. ohms
d. henries

39. For a capacitor to correct the power factor to unity it must have an X_C that equals
a. R
b. Z
c. X_L
d. $\cos \phi$

40. If the starter switch of a switch start fluorescent fitting is removed when the lamp is working the effect will be that the lamp
a. stays working normally
b. blows the circuit fuse
c. will start flashing
d. goes out

41. The normal operating frequency of a high frequency fluorescent circuit is approximately
   a. 50 Hz
   b. 16 kHz
   c. 30 kHz
   d. 50 kHz
42. The symbol shown in the diagram represents a

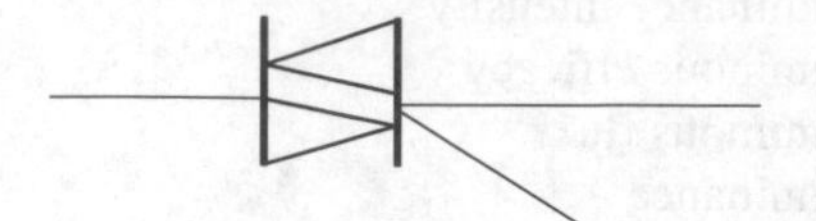

   a. diode
   b. triac
   c. diac
   d. thyristor
43. The auxiliary electrode in the discharge tube of a high pressure mercury vapour lamp shown in the diagram is

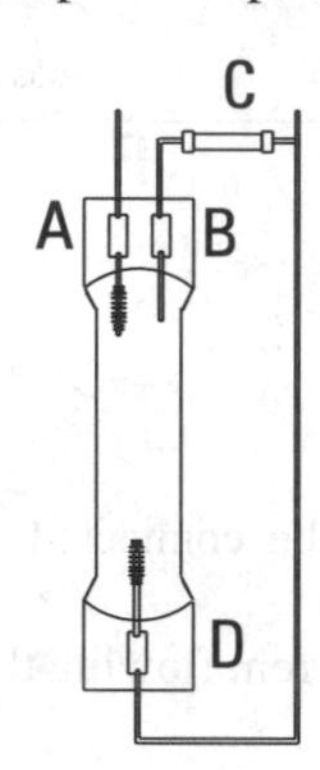

   a. A
   b. B
   c. C
   d. D
44. A continuity test on main equipotential bonding conductors should be carried out with a
   a. bell tester
   b. milliohmmeter
   c. Megohmmeter
   d. earth ohmmeter
45. The permitted insulation test voltage on installations supplied with up to 500 V a.c. r.m.s. is
   a. 250 V d.c.at 0.5 mA
   b. 250 V d.c. at 1.0 mA
   c. 500 V d.c.at 0.5 mA
   d. 500 V d.c.at 1.0 mA
46. The BS3535 shaver socket is an example of protection by
   a. electrical separation
   b. barriers and enclosures
   c. Class 1 equipment
   d. a solid earthed system
47. The resistance between two adjacent heating pipes is found to be 1 MΩ. This means that
   a. the main equipotential bonding is faulty
   b. the two pipes need bonding together
   c. a test should be carried out to earth
   d. the resistance is satisfactory
48. An enclosure which prevents the insertion of wires greater than 1 mm in diameter would be classified as
   a. IP1
   b. IP2
   c. IP3
   d. IP4
49. A polarity test is carried out to confirm that all
   a. switches and single pole control devices are in the phase conductors only
   b. connections are electrically and mechanically sound
   c. socket outlets are connected to a ring final circuit
   d. circuits will function as intended
50. To test that a 30 mA RCD will operate satisfactorily under fault conditions, tests are carried out at 50% and 100% of the rated tripping current and 150 mA. The trip should operate on
   a. only 50%
   b. all three tests
   c. 50% and 100% only
   d. 100% and 150 mA only

Answer grid

1	a	b	c	d	26	a	b	c	d
2	a	b	c	d	27	a	b	c	d
3	a	b	c	d	28	a	b	c	d
4	a	b	c	d	29	a	b	c	d
5	a	b	c	d	30	a	b	c	d
6	a	b	c	d	31	a	b	c	d
7	a	b	c	d	32	a	b	c	d
8	a	b	c	d	33	a	b	c	d
9	a	b	c	d	34	a	b	c	d
10	a	b	c	d	35	a	b	c	d
11	a	b	c	d	36	a	b	c	d
12	a	b	c	d	37	a	b	c	d
13	a	b	c	d	38	a	b	c	d
14	a	b	c	d	39	a	b	c	d
15	a	b	c	d	40	a	b	c	d
16	a	b	c	d	41	a	b	c	d
17	a	b	c	d	42	a	b	c	d
18	a	b	c	d	43	a	b	c	d
19	a	b	c	d	44	a	b	c	d
20	a	b	c	d	45	a	b	c	d
21	a	b	c	d	46	a	b	c	d
22	a	b	c	d	47	a	b	c	d
23	a	b	c	d	48	a	b	c	d
24	a	b	c	d	49	a	b	c	d
25	a	b	c	d	50	a	b	c	d

Appendix

Contents

CT Manufacturing Ltd.

Additional tasks:

Answers:

CT Manufacturing Ltd.

Specification

The drawings show a factory which manufactures small electrical products.

Building construction

The building is of steel frame construction with facing bricks externally and an inner skin of lightweight concrete blocks. The roof over the workshop area and half of the canteen is a double pitched roof supported by steel roof trusses having lattice girders which, in turn, are supported by stanchions positioned as shown in the drawings. The roof is of corrugated sheets with lights fitted over the workshop area. The valley between the double pitch acts as a fire escape from the first floor offices.

The height from the finished floor level to the underside of the horizontal joists is 4.0m.

The floor is a cast-in-situ concrete slab with screed finish. The first floor only covers half of the ground floor area. The roof to this is cast-in-situ concrete slabs covered with standard bitumen to the required thickness.

Electrical installation

The electrical installation is in accordance with the Electricity at Work Regulations 1989 and BS 7671:1992 Requirements for Electrical Installations.

Electricity supply

The supply is three-phase four wire 400/230 V 50 Hz.

The supply and installation form a TN-C-S system protected in the supply company's cut out by 3 × 100 A BSEN 60269-1:1994 (BS88 Part 2 and Part 6) type fuses.

Z_e is 0.3 Ω and I_{pf} is 16 kA.

Wiring

The submain between the two distribution boards is a 35 mm^2 two-core XLPE/SWA/LSF cable with copper conductors which is 28 metres long, run cleated to a perforated steel cable tray and no factors apply. Circuits supplied from the first floor distribution board must not exceed 3.5 volts drop. Measured from the first floor distribution board Z_{db} is 0.8242 Ω and I_s is 5.4 kA. The submain is protected by a 63 A BSEN 60269-1:1994 (BS88 Part 2 and Part 6) type fuse. Power supplies to the main workshop area are taken from an overhead busbar trunking as shown.

Lighting to the ground floor is by PVC insulated single core cables with copper conductors installed in heavy gauge B.E. steel conduit. Ground floor power is via a steel trunking and conduit system using PVC insulated single cables with copper conductors.

First floor circuits are installed in heavy gauge PVC conduit using single core PVC insulated single cables with copper conductors.

Where steel conduit or trunking is used it forms the only cpc for the circuitry.

The power to the boiler room and the whole of the fire alarm system are installed in mims cable with copper conductors and sheath. The cable is served overall with PVC insulation and installed on steel cable tray.

Emergency lighting units are self contained, non-maintained and supplied normally from the appropriate lighting circuit to keep their batteries charged. For the calculation of circuit loads these may be ignored.

Mounting heights for electrical equipment

From F.F.L. to centre

Socket outlets		1 m
Light switches		1.4 m
Kitchen equipment		1.2 m
Distribution boards	ground floor	3 m
	first floor (ceiling height 2.4 m)	1.8 m

Circuit protection

With the exception of the first floor circuits HBC fuses to BSEN 60269-1:1994 are used throughout. Each of the first floor circuits is protected by a Type C MCB to BSEN 60898.

CT Manufacturing Ltd.

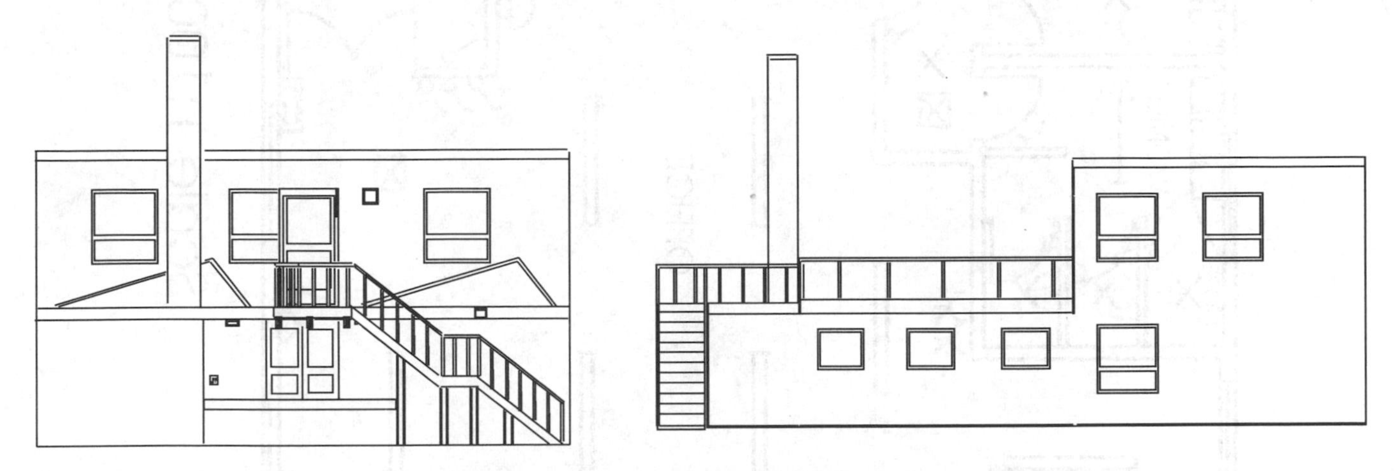

West elevation

South elevation

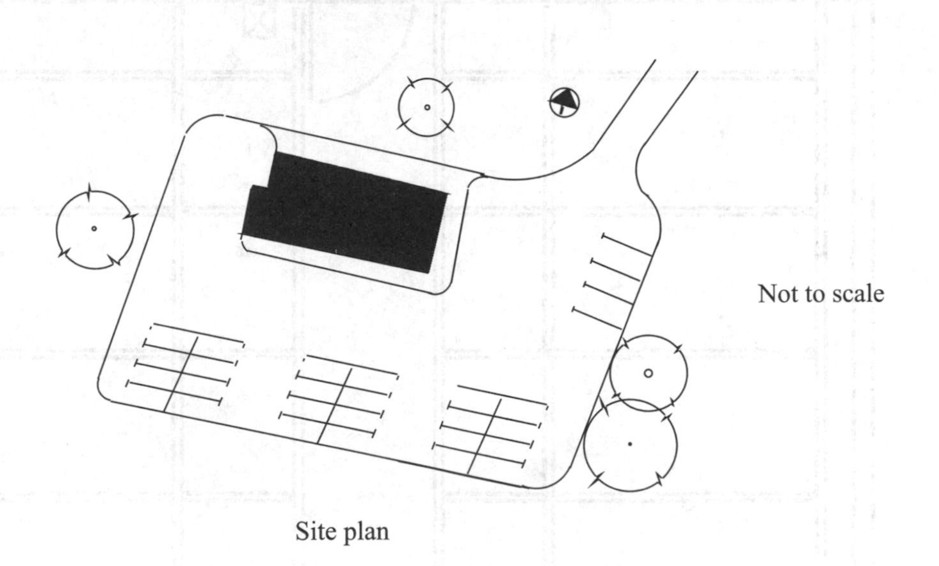

Site plan

- 65 W fluorescent luminaire
- 65 W × 2 fluorescent luminaire
- Tungsten filament luminaire
- Tungsten halogen luminaire (in loading bay)
- Emergency lighting unit
- One way switch
- Two way switch
- 13 A switch socket
- Twin 13 A switch socket
- 20 A switch fused outlet
- Fire alarm call point
- Fire alarm sounder
- Overhead busbar trunking
- Intake

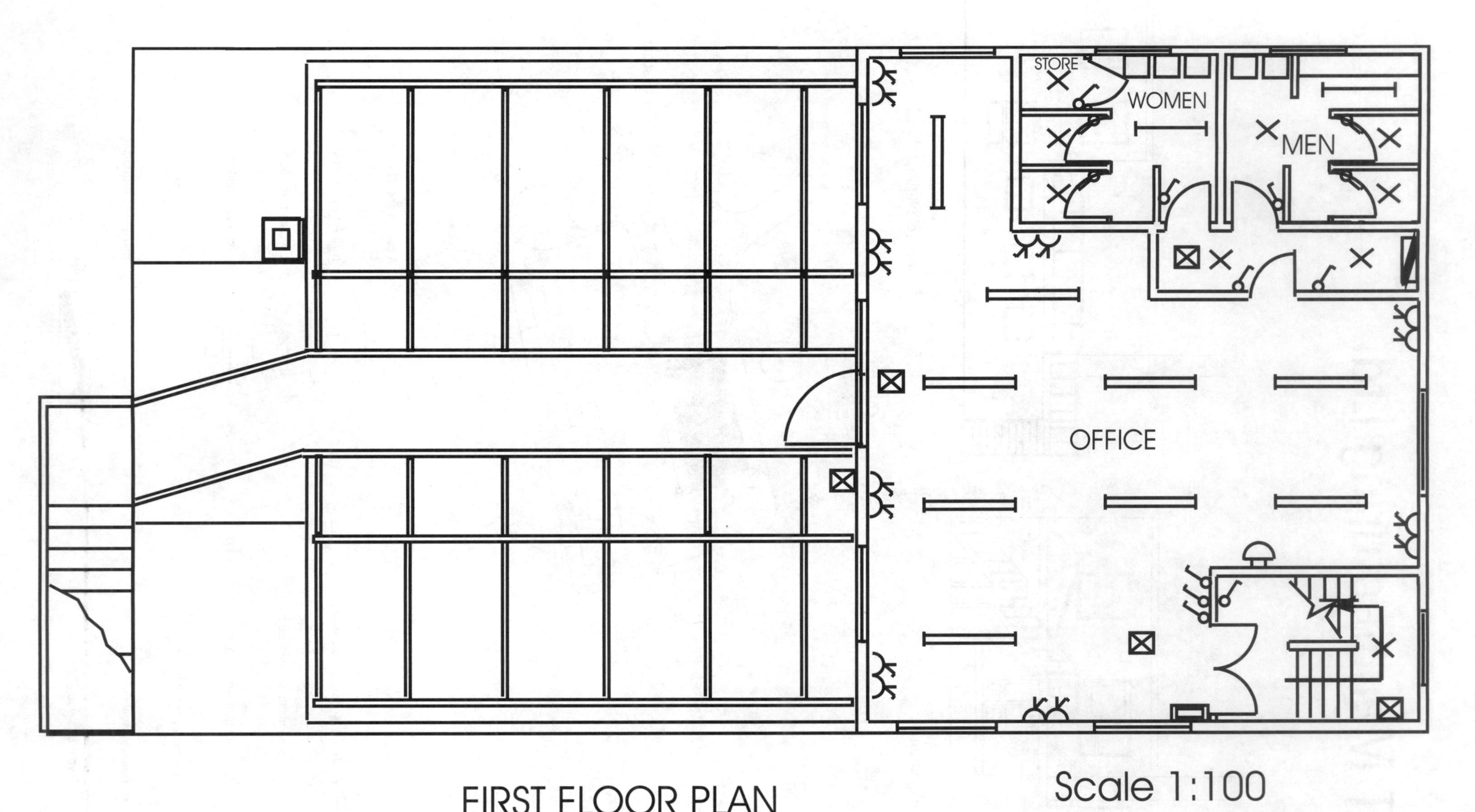
STORE
WOMEN
MEN
OFFICE
FIRST FLOOR PLAN
Scale 1:100

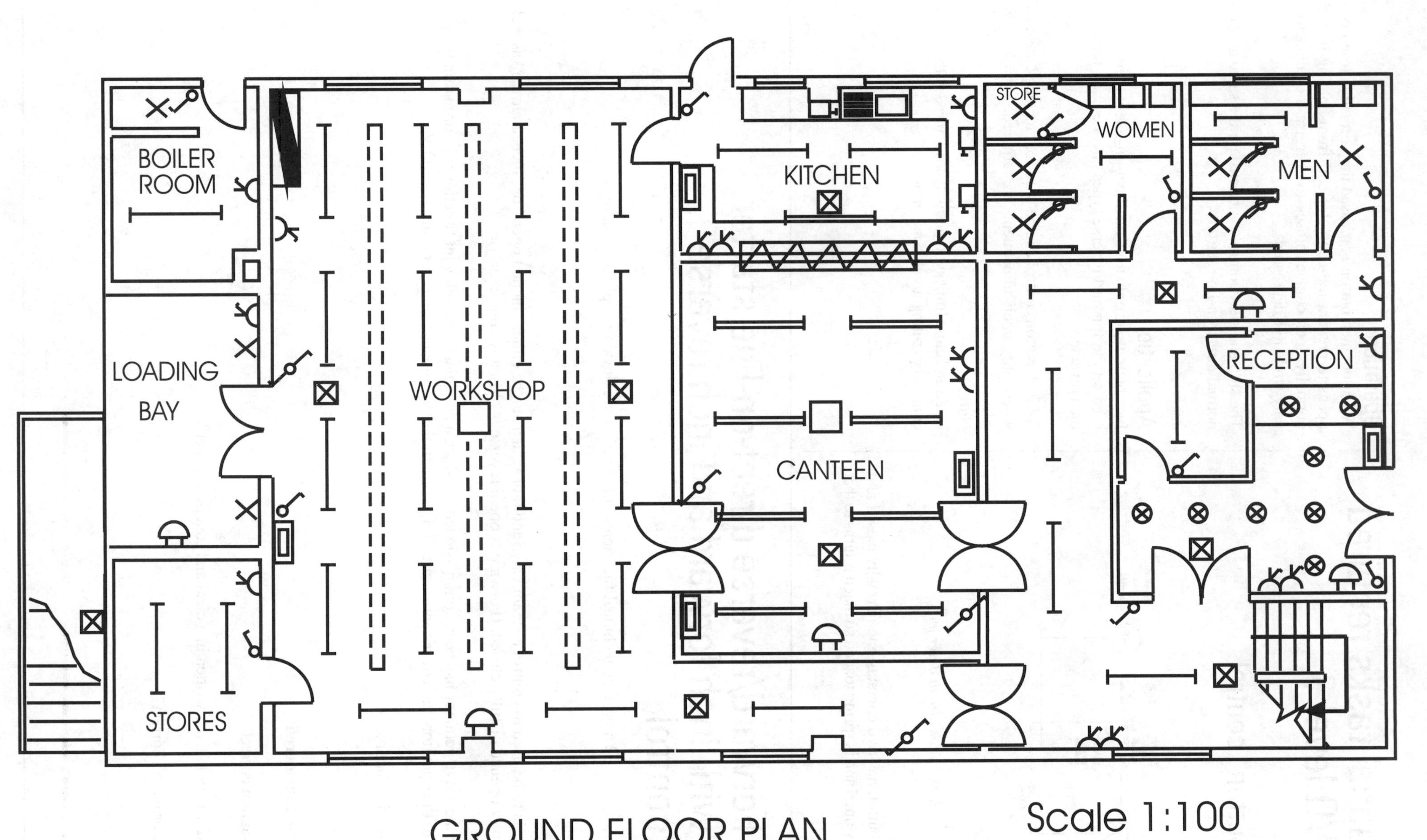

GROUND FLOOR PLAN

Scale 1:100

Additional tasks required for NVQ level 3

"Inch button" control

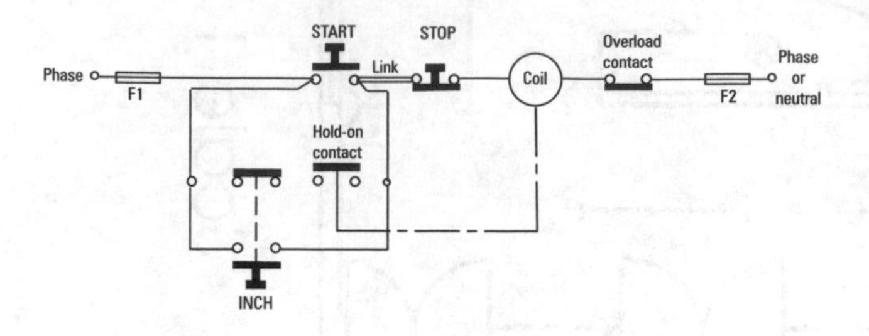

Figure A.1 *Inch button circuit diagram*

The "inch" button has two contacts one normally closed and one normally open and is spring loaded so that it returns to that position.

Operation

When the inch button is pressed the normally closed contact opens to prevent the contactor coil from holding in and the normally open contact closes to energise the contactor coil only while it is depressed.

The stop and start control functions are the same as with the normal direct-on-line starter.

Application

Inch buttons are only used on direct-on-line starters and can be used on the forward/reverse types also i.e. inch forward and inch reverse.

An inch button control can be used for "inching" open valves where only small adjustments are required.

Note:
Open and close limit switches will be required with this type of control operating a valve actuator.

46 Forward/reverse direct-on-line starter with inch forward and inch reverse control

Objective: To provide technical functional information to relevant people on handover.

Note:

The starter is to be used to control a steam valve actuator which can be either "inched open" or "inched closed" as well as being fully opened or fully closed. (Forward is open, reverse is closed for this exercise.)

Assuming the valve actuator has been commissioned, produce a document with relevant functional information for handover to the operator. Include a control circuit diagram in the document. Contactor coils are 400 V.

Time allowed: 2 hours

Points to be considered:

✓ is the diagram correct?

✓ is the information given written in a clear and concise manner?

✓ is the information correct?

47 Luminaire Installations

Objective: To install: (a) a tungsten halogen luminaire on an outside brick wall using a double extension ladder for access.
(b) a high pressure mercury vapour luminaire to an overhead rolled steel joist (or similar) using a tower scaffold for access.

Note: Each luminaire will be one-way controlled and the wiring system is left to the tutor's discretion.

SAFE ELECTRICAL SITE WORKING IS OF PARAMOUNT IMPORTANCE AT ALL TIMES.

Time allowed: for each installation depends on the type of wiring systems used.

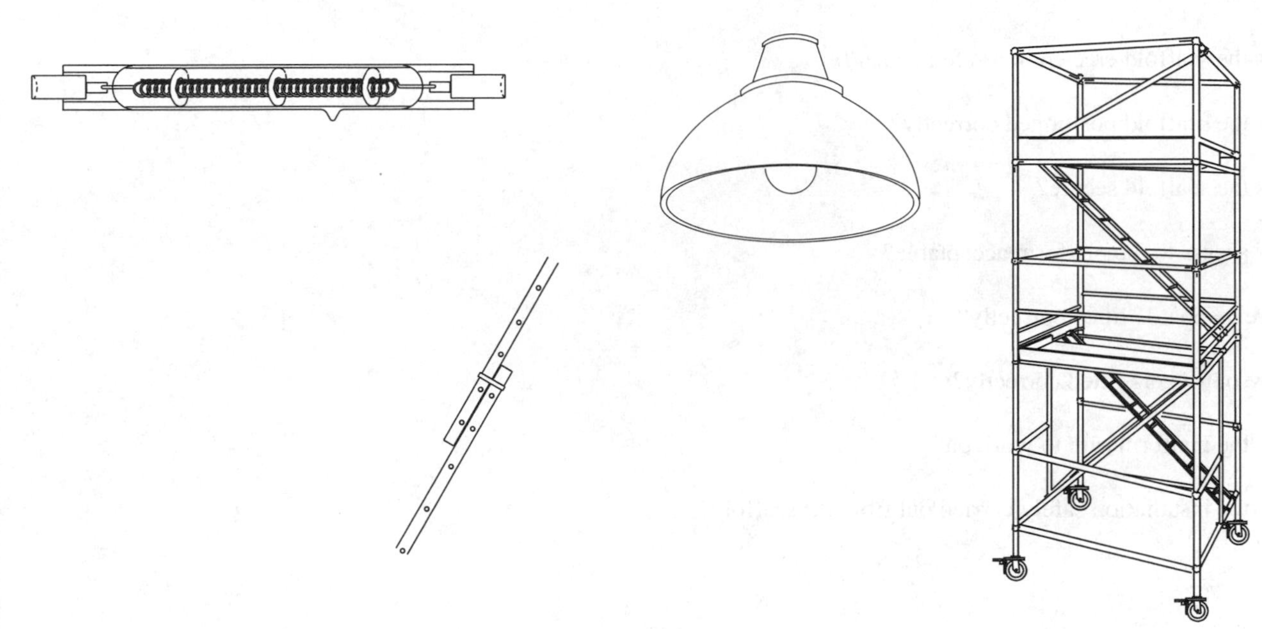

Figure A.2

Points to be considered:

Electrical installation:

✓ are the luminaires installed correctly?

✓ has the wiring been carried out correctly?

✓ are all terminations secure?

✓ are all terminations shielded from direct contact?

✓ have all enclosures been securely fastened?

✓ have the installations been carried out in a safe manner?

✓ have the installations been inspected and tested?

✓ do the circuits function correctly?

Ladder:

✓ was the ladder lifted and carried safely?

✓ was the ladder positioned correctly?

✓ was the ladder ascended in a safe manner?

✓ was the ladder secured safely?

✓ was the installation work carried out safely from the ladder?

Scaffold:

✓ was the scaffold erected in a safe manner?

✓ was the scaffold positioned correctly?

✓ was the scaffold secure?

✓ was access to the platform acceptable?

✓ were toeboards fitted correctly?

✓ were outriggers fitted correctly?

✓ was the platform safe to work on?

✓ was the installation safely carried out from the scaffold?

Answers

These answers are given for guidance and are not necessarily the only possible solutions.

Chapters 1 to 6

p.3 Try this: 1. 522-08-03; 2. 514-13-01 (ii); 3. 602-07-01; 4. 412-06-01
p.32 flameproof accessories, terminations etc.
p.35 Try this: 2.5 mm^2 (2.925 V)
p.37 Try this: (a) 0.63 Ω (b) 365 A
p.38 in series on diagram
p.68 see 10.1 page 62 On Site Guide
p.69 see page 53 in this book
p.71 refer to IP Codes
p.94 refer to Regulation 712-01-03
p.95 0.8 lagging

Chapter 8 – Revision exercises Section one

p.104 Try this: (a) using Figure 7.1, 230 V loads, R to N, Y to N, B to N; 400 V, Y to B (b) 4.3 kW
p.105 Try this:

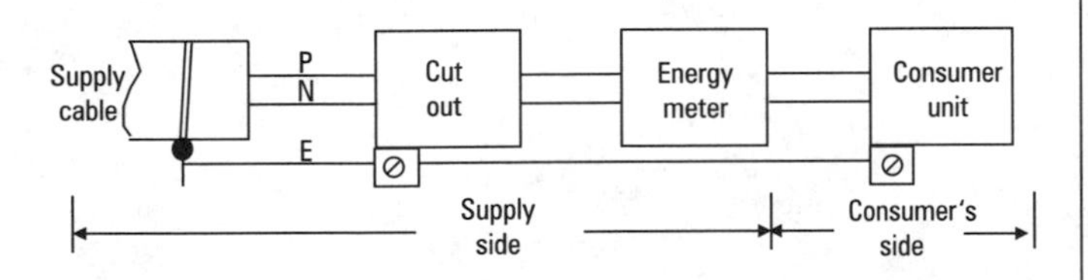

p.106 Try this: (a) none (b) remains the same (c) increases
p.106 Try this: change any two phases
p.108 Try this:

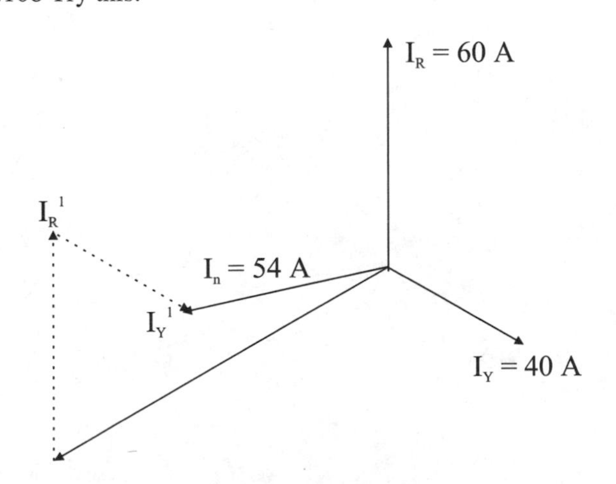

p.108 Try this: (a) to prevent them tripping during starting (b) oil dashpot Figure 4.11

Chapter 8 – Revision Exercises Section two

Supply systems

1. 0.54
2. 34.6 A
3. combined on supply side, separate on consumer's side
4. disconnector
5. easier to tap-off at each floor level
6.

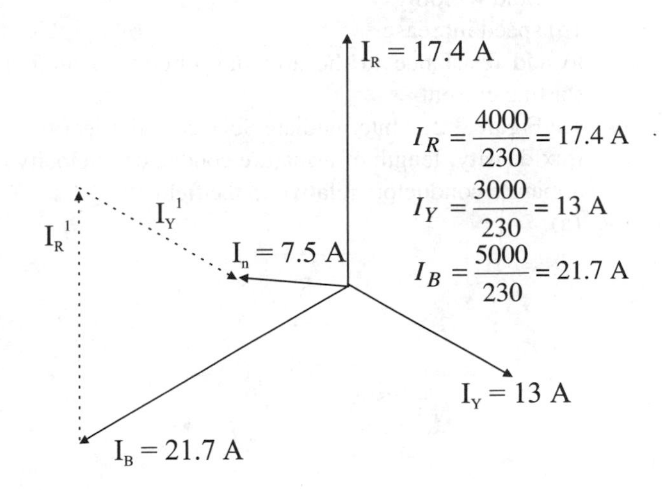

7.

230 V a.c. supply — 18 V — C.T. — D_1 — D_2 — 9 V d.c. load

Wiring systems

1. (a) PVC/PVC in mini trunking, easy to install (b) mims fire resistant (c) mims fire resistant
2. 25 mm^2
3. (a) 0.42 Ω (1.2 multiplier used)
   (b) 0.72 Ω
   (c) 1.48 Ω
4. (a) 980 A (b) 0.6s
5. see p.116, "Stage 1 Design"

A.C. motors

1. (a) change armature or field, not both (b) any two phases to stator change
2. cooling fan circulating air through motor and between fins on stator frame
3. when the motor temperature increases, the P.T.C. type thermistor's resistance increases rapidly, this in turn reduces the control current and the contactor coil de-energises.
4. 10.87 A
5. (a) 15 Ω, (b) 15.33 A, (c) 9.58 A, (d) 9.58 A
6. 24 A
7. at 1450 rev/min = 3.33%, at 2800 rev/min = 6.67%

D.C. machines

1. (a) variable resistor (field regulator) in series with the field windings
   (b) speed increases
2. to add resistance in the armature circuit, to limit the starting current
3. see Figure 3.21 "Intermediate Science and Theory"
4. flux density, length of armature conductors, velocity of armature conductors relative to the field
5. (a)

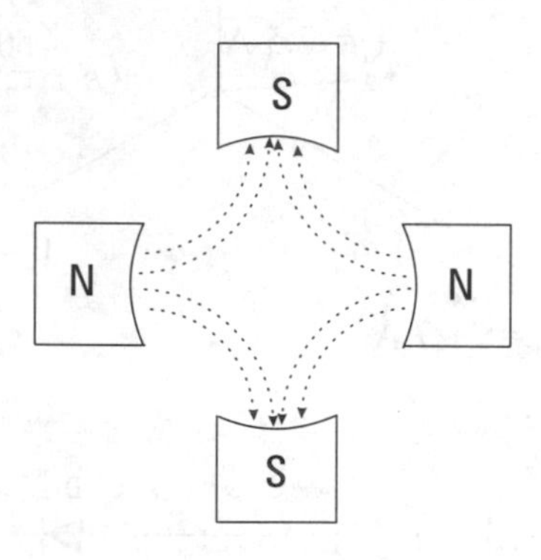

   (b) d.c. motor or d.c. generator

Transformers

1. (a) copper and iron losses
   (b) iron
2. (a) 20 V
   (b) 100 A
3.

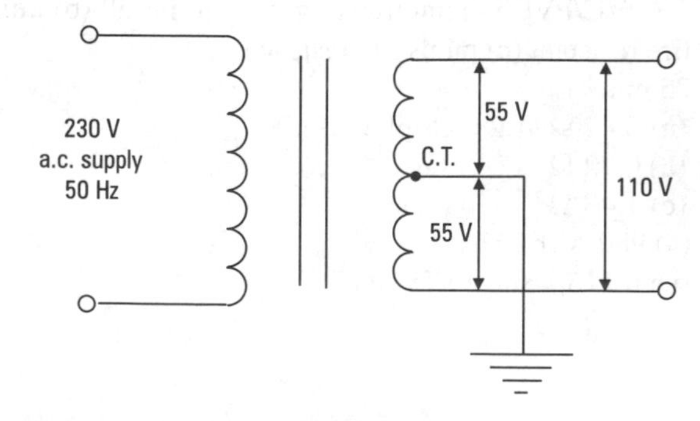

4. 0.26 A
5. 92%

Lighting

1. (a) Figure 8.44 "Intermediate Science and Theory"
   (b) p.161 "Intermediate Science and Theory"
2. (a) Current taken by the control gear must also be included (b) 6 circuits
3. (a) lamp flicker makes rotating machinery to appear to be stationary when it is not (b) the tube remains on
4. (a) 50 lux (b) 25.6 lux
5. 20,000 lumens

Inspection and Testing

1. (a) poor discrimination, may not operate (b) cannot handle short circuit currents very well, may suffer damage
2. p.86 "Stage 1 Design"
3. using proprietary test leads, one on the phase the other on the earth. Test at motor terminal box.
4. p.97 "Stage 1 Design"
5. p.88 "Stage 1 Design"

Multi-choice answers

1. c; 2. b; 3. b; 4. d; 5. a; 6. b; 7. c; 8. c; 9. b; 10. d; 11. b; 12. b; 13. d; 14. c; 15. d; 16. d; 17. b; 18. d; 19. a; 20. d; 21. d; 22. b; 23. c; 24. b; 25. c; 26. d; 27. a; 28. b; 29. c; 30. c; 31. d; 32. a; 33. c; 34. b; 35. b; 36. b; 37. a; 38. c; 39. c; 40. a; 41. c; 42. b; 43. b; 44. b; 45. d; 46. a; 47. b; 48. d; 49. a; 50. d